ISBN 978-0-265-90528-9
PIBN 10907597

This book is a reproduction of an important historical work. Forgotten Books uses
state-of-the-art technology to digitally reconstruct the work, preserving the original format
whilst repairing imperfections present in the aged copy. In rare cases, an imperfection in
the original, such as a blemish or missing page, may be replicated in our edition. We do,
however, repair the vast majority of imperfections successfully; any imperfections that
remain are intentionally left to preserve the state of such historical works.

TECHNICAL BULLETIN No. 32. ~}45   JANUARY, 1914.

New York Agricultural Experiment Station.

GENEVA, N. Y.

A CONTRIBUTION TO THE CHEMISTRY OF PHYTIN.

I. THE ORGANIC PHOSPHORIC ACID OF COTTONSEED MEAL, II.

II. PHYTIN IN OATS.

III. PHYTIN IN CORN.

IV. CONCERNING THE COMPOSITION OF BARIUM PHYTATE AND PHYTIC
ACID FROM COMMERCIAL PHYTIN AND A STUDY OF THE PROP-
ERTIES OF PHYTIC ACID AND ITS DECOMPOSITION PRODUCTS.

R. J. ANDERSON.

PUBLISHED BY THE DEPARTMENT OF AGRICULTURE.

BOARD OF CONTROL.

TECHNICAL BULLETIN No. 32.

A CONTRIBUTION TO THE CHEMISTRY OF PHYTIN.

SUMMARY.

R. J. ANDERSON.

This bulletin contains the report of an investigation concerning the composition of the organic phosphoric acids of cottonseed meal, oats and corn in comparison with commercial phytin.

It is shown that from all of these substances identical barium salts are obtained which agree very closely in composition with the following types of salts, viz:

Tribarium inosite hexaphosphate, $C_6H_{12}O_{24}P_6Ba_3$, obtained as minute bundles or globules of microscopic needles from dilute hydrochloric acid solutions by the addition of alcohol, and hepta-barium inosite hexaphosphate $(C_6H_{11}O_{24}P_6)_2Ba_7$, or $C_{12}H_{22}O_{48}P_{12}Ba_7$, which separates from dilute hydrochloric acid solutions in the presence of barium chloride in globular masses of needle-shaped crystals.

The free acid prepared from the crystalline barium salts agrees more closely in composition with inosite hexaphosphate, $C_6H_{18}O_{24}P_6$, than with the usual formula for phytic acid, $C_6H_{24}O_{27}P_6$.

Oats apparently contains two different organic phosphoric acids but only one, that corresponding to inosite hexaphosphate, has been isolated in pure form.

The spontaneous decomposition products of phytic acid under ordinary conditions which are formed within a reasonable length of time appear to be phosphoric acid and substances which contain more carbon and less phosphorus than phytic acid, which substances are probably penta-, tetra-, etc., phosphoric acid esters of inosite.

When phytic acid is dried at a temperature of 105° under reduced pressure, it rapidly decomposes with liberation of inorganic phosphoric acid and the formation of various decomposition products, consisting of inosite and substances varying in composition from inosite tetraphosphate to inosite monophosphate.

When the crystalline barium salts are dried at 105° under reduced pressure they suffer but slight hydrolysis. Under ordinary conditions the dry salts are comparatively stable but on longer keeping small quantities of inorganic phosphoric acid are liberated.

From the analytical data reported it appears that the substance known as phytic acid or inosite phosphoric acid is either inosite hexaphosphate, $C_6H_{18}O_{24}P_6$ or else an isomer and that the formulas, $C_2H_8O_9P_2$ or $C_6H_{27}O_{18}P_6$ heretofore used to represent this acid are incorrect.

CONCERNING THE ORGANIC PHOSPHORIC ACID OF COTTONSEED MEAL. II.

In the last report[1] from this laboratory we described certain crystalline barium salts of the organic phosphoric acid of cotton-seed meal. We had also prepared and analyzed the free acid itself and described its properties and we also showed that on cleavage with dilute sulphuric acid in a sealed tube the substance gave inosite as one of the products of decomposition.

These crystalline barium salts and the free acid prepared from them gave results on analysis which differed slightly from corresponding compounds calculated on the usual formula for phytic acid, viz: $C_6H_{24}O_{27}P_6$. The substance from cottonseed meal appeared to be an acid of the formula $C_2H_6O_8P_2$ or $C_6H_{18}O_{24}P_6$. The barium salts agreed closely with this formula but the percentage of phosphorus in the free acid was found to be about 1 per ct. lower than required.

The reactions of the aqueous solution of the free acid, however, were found to be identical in every respect with those given by phytic acid. From the results obtained we concluded that the organic phosphoric acid in cottonseed meal was very similar to phytic acid but we were unable to determine whether it was identical with this acid.

Prior to our publication, so far as we are aware, no definite organic phosphoric acid had ever been described as existing in cottonseed meal; no pure salts of this acid had been obtained nor had the free acid been prepared in pure form.

However, some earlier work had been published by Rather[2] dealing with " The forms of phosphorus in cottonseed meal." This author had isolated certain more or less impure substances from cottonseed meal which undoubtedly contained some of the organic phosphoric acid which we later isolated in pure form. He found that these preparations gave reactions similar to those of meta- and pyrophosphoric acids and he concluded that these reactions therefore were not sufficient to prove that either meta- or pyrophos-phoric acid exists in cottonseed meal as had been claimed earlier.[3]

[1] *Journ. Biol. Chem.* 13:311, 1912, and N. Y. Agr. Exp. Sta. Tech. Bull. 25, 1912.
[2] Texas Agr. Exp. Sta., Bull. 146.
[3] Hardin. S. C. Agr. Exp. Sta., Bull. 8, N. S., 1892.

From the acid preparations which he had isolated, he prepared some silver salts for which he proposed the following formulas, viz:

Product A, $C_4H_{12}Ag_5P_3O_{15}$.
Product B, $C_6H_{10}Ag_7P_4O_{17}$.
Product C, $C_4H_{10}Ag_5P_3O_{13}$.

In a more recent publication by the same author [4] are reported the analyses of a few more amorphous silver salts prepared, by a method similar to the one used before, from cottonseed meal and wheat bran. It is claimed that these compounds are identical, i.e., they are salts of the same acid, as shown by their having the same percentage composition, the same solubility, etc. This time, however, these amorphous compounds are alleged to be salts of an acid of the formula $C_{12}H_{41}P_9O_{42}$ and which formula is proposed as the correct one for the substance known as inosite phosphoric acid or phytic acid.

Since these results did not harmonize with our earlier findings in respect to cottonseed meal Mr. Rather suggests that the carefully purified and recrystallized barium salts which we had analyzed must have contained " iron, aluminum, lime and magnesia "— this being the more likely since we had presented no analytical data to show that these inorganic substances were absent. Evidently Mr. Rather had not read our publication very carefully, otherwise he might have noticed that we stated, concerning the barium salts, that, " metals other than barium were absent." [5]

In the present paper we wish to refer to the work on cottonseed meal only, reserving for a later communication proofs to show that the results reported by Rather are just as inapplicable to the organic phosphorus compound of wheat bran as they are to the acid existing in cottonseed meal.

Since our earlier work had shown that the organic phosphoric acid of cottonseed meal gave barium salts which crystallized readily and which could be easily purified by repeated recrystallizations and since it is generally recognized that crystalline substances are more suitable for the identification of chemical compounds than amorphous bodies we have repeated our former work on cottonseed meal in the hope of establishing more definitely the composition of the organic phosphoric acid present in this material.

From 25 pounds of cottonseed meal we obtained, after recrystallizing eleven times, 69 grams of the barium salt. So far as composition, crystal-form and reactions are concerned this product was identical with the salts previously described. Further recrystal-

[4] *J. Am. Chem. Soc.* 35:890, 1913, and Texas Agr. Exp. Sta., Bull. 156, 1913.
[5] *Loc. cit*, p. 321 and p. 11.

lizations did not alter the composition. Heavy metals other than barium were absent and we could not detect any weighable quantity of alkalies in 0.5 gram of the salt. It was completely free from inorganic phosphate and it was free from nitrogen and sulphur. We believe, therefore, that it represénts a pure chemical compound.

The composition as previously reported [6] agrees very closely with that required by inosite hexaphosphate, $C_6H_{18}O_{24}P_6$. The free acid was prepared and analyzed, which also agreed with the above formula. Silver salts were prepared from the above isolated acid but it would seem that silver salts are not very suitable for the purpose of identifying an acid of the above nature. They are obtained as amorphous precipitates which do not represent homogeneous salts. They are evidently mixtures of more or less acid silver salts.

The silver precipitates which we obtained did not agree in composition with the compounds analyzed by Rather nor did they agree with any definite silver salts of inosite hexaphosphate. As suggested above, they are evidently mixtures of more or less acid silver salts of inosite hexaphosphate — for after deducting the amount of silver found, allowing for a corresponding amount of hydrogen and water and calculating to the free acid, the results agree very closely with the percentage composition calculated for inosite hexaphosphate.

In the isolation and purification of the barium salt we made use of our former method in preference to that proposed by Rather for the reason that we consider our method more simple and convenient. The essential difference in these methods of isolation is that we use barium hydroxide throughout, precipitating the substance with this reagent from dilute hydrochloric acid solutions. Rather used a modification of the method of Patten & Hart [7] — substituting the use of sodium hydroxide with ammonium hydroxide. The use of either sodium or ammonium hydroxide which must be eliminated again is not necessary, for barium hydroxide is equally efficient and by its use the introduction of other basic ions is avoided.

Since the present work substantiates our earlier results and since all the analytical data agrees with inosite hexaphosphate, $C_6H_{18}O_{24}P_6$, or with salts of this acid, we believe that the organic phosphoric acid in cottonseed meal must be represented by the formula either of inosite hexaphosphate, $C_6H_{18}O_{24}P_6$, or else some formula isomeric with this.

It may be noted that the percentage of phosphorus found on analyzing the free acid is somewhat low. In the analyses of the acids previously reported [8] the phosphorus was found to be from

---

[6] *Loc. cit.*

[7] *Am. Chem., Journ.* 31:566, 1904.

[8] *Loc. cit.*

1 to 1.8 per ct. lower than required for inosite hexaphosphate. As will be shown later in this bulletin, this is due to 'the fact that the free acid becomes largely hydrolyzed on drying.

EXPERIMENTAL PART.

ISOLATION AND PURIFICATION OF THE BARIUM SALT.

The cottonseed meal, 25 pounds, was digested over night in 0.2 per ct. hydrochloric acid in porcelain percolators covered on the inside with a double layer of cheesecloth. It was then percolated using 0.2 per ct. hydrochloric acid until about 20 liters of extract were obtained. The extract was of a dirty, dark color and contained some suspended particles from which it was freed as much as possible by centrifugalizing the solution. A concentrated solution of 300 grams of barium chloride was then added and the precipitate allowed to settle. The precipitate was centrifugalized and finally brought upon a Buchner funnel and freed as far as possible from the mother-liquor. It was then digested in several liters of about 5 per ct. hydrochloric acid until no further solution took place. The insoluble residue was removed by centrifugalizing and the still very dirty colored solution precipitated by adding barium hydroxide until the free acid was neutralized. The barium hydroxide was added slowly, with constant shaking, when the precipitate separated in crystalline form. It was then filtered and washed thoroughly in water and again dissolved in dilute hydrochloric acid, filtered and reprecipitated with barium hydroxide. These operations were repeated three times. The hydrochloric acid solution was then precipitated by gradually adding an equal volume of alcohol when the substance again separated in crystalline form consisting of globular masses of microscopic needles. It was then precipitated a fourth time with barium hydroxide and after that two more times with alcohol. It was then filtered, washed free of chlorides with dilute alcohol and then in alcohol and ether and dried in vacuum over sulphuric acid. The product was then a nearly white, crystalline powder and it weighed 94 grams.

The dry substance was shaken up with about 1.5 liters of cold water, allowed to stand for several hours and then filtered and washed in water. The aqueous solution contained very little substance precipitable with alcohol and it was therefore discarded.

The washed precipitate was dissolved in dilute hydrochloric acid and precipitated a fifth time by the very gradual addition of barium hydroxide; after filtering and washing, this operation was repeated a sixth time. After again dissolving in dilute hydrochloric acid, nearly neutralizing the free acid with barium hydroxide and filtering, the substance was brought to crystallization by the gradual addition of an equal volume of alcohol. After standing for several hours the substance was filtered and washed in dilute alcohol,

alcohol and ether and dried in vacuum over sulphuric acid. It was then a voluminous snow-white crystalline powder.

The dry substance was again dissolved in dilute hydrochloric acid, the free acid nearly neutralized with barium hydroxide and the solution filtered and allowed to stand over night. The substance soon began to crystallize. Under the microscope it appeared perfectly homogeneous and consisted as before of globular masses of microscopic needles. The substance was filtered, washed free of chlorides with water and then in alcohol and ether and dried in vacuum over sulphuric acid. The dry, snow-white, crystalline powder weighed 69 grams.

Qualitative analysis failed to reveal any heavy metals other than barium and from 0.5 gram of the salt no weighable residue of alkali was obtained. It gave no reaction with ammonium molybdate in nitric acid solution. It was free from sulphur and nitrogen.

It was analyzed after drying at 105° in vacuum over phosphorus pentoxide.

0.4641 gram subst. gave 0.0556 gram H_2O and 0.1125 gram CO_2.
0.1982 gram subst. gave 0.1333 gram $BaSO_4$ and 0.1203 gram $Mg_2P_2O_7$.

Found: C = 6.61; H = 1.34; P = 16.91; Ba = 39.57 per ct.

For tribarium inosite hexaphosphate:
$C_6H_{12}O_{24}P_6Ba_3$ = 1066.
Calculated: C = 6.75; H = 1.12; P = 17.44; Ba = 38.65 per ct.

A portion of this salt was recrystallized as follows: 5 grams were dissolved in a small quantity of 5-per-ct. hydrochloric acid and the free acid nearly neutralized with barium hydroxide, the solution was then filtered and 2 grams of barium chloride dissolved in a little water added and the solution allowed to stand. The substance separated slowly in the usual crystal form. After two days it was filtered, washed free of chlorides with water, again dissolved in the dilute hydrochloric acid, the solution filtered and alcohol added gradually until a slight cloudiness remained. After standing for 24 hours at room temperature the substance had crystallized in the usual form. It was filtered, washed free of chlorides in dilute alcohol and then in alcohol and ether, and dried in vacuum over sulphuric acid. The dilute nitric acid solution of the substance gave no reaction with ammonium molybdate. The snow-white crystalline powder was analyzed after drying at 105° in vacuum over phosphorus pentoxide.

0.3588 gram subst. gave 0.0464 gram .H_2O and 0.0867 gram CO_2.
0.1726 gram subst. gave 0.1138 gram $BaSO_4$ and 0.1058 gram $Mg_2P_2O_7$.

Found: C = 6.59; H = 1.44; P = 17.08; Ba = 38.79 per ct.

Another portion of the substance was recrystallized as follows: 2 grams were dissolved in a small amount of the dilute hydrochloric acid, barium hydroxide added, with constant shaking, until a faint permanent precipitate remained, and the solution filtered. The filtrate was then heated to boiling and allowed to stand for a few minutes. As the temperature rose the solution began to turn cloudy and finally a heavy precipitate separated which appeared to be amorphous at first but it soon changed into the same crystal form as previously described. This was filtered and washed free of chlorides in boiling water and then in alcohol and ether and allowed to dry in the air. The dry substance weighed 1.6 grams. The snow-white crystalline powder was free from inorganic phosphate. It was analyzed after drying at 105° in vacuum over phosphorus pentoxide.

0.4695 gram subst. lost 0.0443 gram H_2O.
0.4252 gram subst. gave 0.0423 gram H_2O and 0.0981 gram CO_2.
0.1238 gram subst. gave 0.0885 gram $BaSO_4$ and 0.0735 gram $Mg_2P_2O_7$.

Found: $C = 6.29$; $H = 1.11$; $P = 16.54$; $Ba = 42.06$; $H_2O = 9.43$ per ct.

For heptabarium inosite hexaphosphate $(C_6H_{11}O_{24}P_6)_2Ba_7 = 2267$. Calculated: $C = 6.35$; $H = 0.97$; $P = 16.40$; $Ba = 42.39$ per ct. For 14 H_2O calculated, 10.00 per ct.

Still another portion of the substance was recrystallized in the following manner: 2 grams were dissolved in the dilute hydrochloric acid and then nearly neutralized with barium hydroxide as before. The solution was filtered and 10 c.c. $^N/_1$ barium chloride added and allowed to stand over night. The substance had then separated as a heavy crystalline powder of the same form as before except that the individual crystals were much larger. The crystals were filtered, washed free of chlorides with water and finally in alcohol and ether and allowed to dry in the air. It was analyzed after drying as above.

0.6430 gram subst. lost 0.0745 gram H_2O.
0.5685 gram subst. gave 0.0603 gram H_2O and 0.1258 gram CO_2.
0.2208 gram subst. gave 0.1608 gram $BaSO_4$ and 0.1252 gram $Mg_2P_2O_7$.

Found: $C = 6.03$; $H = 1.18$; $P = 15.80$; $Ba = 42.85$; $H_2O = 11.58$ per ct.

For heptabarium inosite hexaphosphate $(C_6H_{11}O_{24}P_6)_2Ba_7 = 2267$.

Calculated: $C = 6.35$; $H = 0.97$; $P = 16.40$; $Ba = 42.39$ per ct. For 16 H_2O calculated: 11.27 per ct.

PREPARATION OF THE FREE ACID

The acid was prepared in the usual way from 10 grams of the first crystalline barium salt. The aqueous solution finally obtained was concentrated in vacuum at 40° to 45° to small bulk. It was then divided into three portions — one was dried in vacuum over sulphuric acid and then analyzed — the others were used for the preparation of the silver salts to be described later.

The dry acid was obtained as a practically colorless syrup. Its dilute aqueous solution gave no reaction with ammonium molybdate showing absence of inorganic phosphoric acid. Its reactions in other respects were identical with those previously described. For analysis it was dried first in vacuum over sulphuric acid at room temperature and finally in vacuum over phosphorus pentoxide at 105°, when it turned quite dark in color.

0.3931 gram subst. gave 0.1088 gram H_2O and 0.1540 gram CO_2.
0.1840 gram subst. gave 0.1826 gram $Mg_2P_2O_7$.

Found: $C = 10.68$; $H = 3.09$; $P = 27.66$ per ct.
For inosite hexaphosphate, $C_6H_{18}O_{24}P_6 = 660$.
Calculated: $C = 10.90$; $H = 2.72$; $P = 28.18$ per ct.

PREPARATION OF THE SILVER SALT FROM THE ABOVE ACID.

One portion of the free acid previously mentioned was dissolved in 100 c.c. of water and the solution neutralized to litmus with ammonia. Silver nitrate solution was then added which caused a heavy, perfectly white, amorphous precipitate. This was filtered and carefully washed in water and dried in vacuum over sulphuric acid, desiccator being kept in a dark place. After drying, the substance was a faintly cream-colored powder which was very slightly sensitive to light. On moist litmus paper it showed a strong acid reaction. It was free from ammonia. For analysis it was dried at 105° in vacuum over phosphorus pentoxide. On drying as above, not protected from light, the substance darkened somewhat in color.

0.3064 gram subst. gave 0.0150 gram H_2O and 0.0446 gram CO_2.
0.1640 gram subst. gave 0.1387 gram AgCl and 0.0596 gram $Mg_2P_2O_7$.

Found: $C = 3.96$; $H = 0.54$; $P = 10.13$; $Ag = 63.65$ per ct.

Deducting the above percentage of silver and allowing for an equivalent amount of hydrogen and water we obtain the following results:

Calculated: $C = 10.74$; $H = 3.08$; $P = 27.45$ per ct.

These percentages agree fairly closely with the composition calculated for inosite hexaphosphate, viz:

$C = 10.90$; $H = 2.72$; $P = 28.18$ per ct.

To the remaining portion of the acid (about 5 c.c.) 300 c.c. of alcohol was added.* The solution remained perfectly clear. The alcohol was evaporated on the water-bath and the residue taken up in 100 c.c. of water in which it gave a slightly cloudy solution and which had a faint, aromatic odor. The acid had possibly been esterified to a slight extent. It was filtered and neutralized to litmus with ammonia and precipitated with silver nitrate; the precipitate filtered, washed in water and dried as before. The appearance of the precipitate was identical with the first one. On moist litmus paper it also showed a strong acid reaction and it was free from ammonia. For analysis it was dried as above.

0.4008 gram subst. gave 0.0176 gram H_2O and 0.0600 gram CO_2.
0.1481 gram subst. gave 0.1241 gram AgCl and 0.0548 gram $Mg_2P_2O_7$.

Found: $C = 4.08$; $H = 0.49$; $P = 10.31$; $Ag = 63.06$ per ct.

Calculated to the free acid as before the following percentages are obtained:

$C = 10.88$; $H = 2.84$; $P = 27.52$ per ct.

That the silver precipitates obtained under above conditions do not represent homogeneous salts may be seen by comparing the percentages found with the calculated composition of the following silver salts of inosite hexaphosphate:

$C_6H_7O_{24}P_6Ag_{11} = 1836$.
Calculated: $C = 3.92$; $H = 0.38$; $P = 10.13$; $Ag = 64.65$ per ct.
$C_6H_8O_{24}P_6Ag_{10} = 1729$.
Calculated: $C = 4.16$; $H = 0.45$; $P = 10.75$; $Ag = 62.40$ per ct.

Judging by the analytical results the amorphous silver precipitates appear to be mixtures of the above silver salts.

---

*Note: Mr. Rather found that his acid preparations gave a precipitate on addition of alcohol. This was no doubt due to the fact that inorganic bases had not been completely removed by his method of purification; hence an acid salt of the organic phosphoric acid was precipitated on the addition of alcohol. The acid prepared from our purified and recrystallized barium salts is completely soluble in alcohol.

CONCERNING PHYTIN IN OATS.

In continuation of the investigation of the organic phosphoric acids of grains and feeding materials which has been carried out in this laboratory we have examined recently the compound existing in oats. This substance has already been studied by other investigators, notably by Hart and Tottingham [1] who came to the conclusion that oats contained phytin. The purpose of the present investigation was to determine whether the phytin in oats was identical with other phytin preparations obtained from other grains.

We have previously shown that cottonseed meal [2] contains an organic phosphoric acid which differs slightly in composition from that required for the phytic acid formula of Posternak, viz: $C_2H_8O_9P_2$ or according to Neuberg $C_6H_{24}O_{27}P_6$, although so far as properties and reactions were concerned no differences could be observed. This acid from cottonseed meal had been isolated as a crystalline barium salt and since this salt did not show any change in composition on recrystallization we felt reasonably certain that it was a homogeneous substance. On the other hand we were unable to obtain any crystalline barium salts of the organic phosphoric acid of wheat bran.[3] Only amorphous salts were obtained, which differed entirely in composition from salts of phytic acid. It appeared of interest, therefore, to determine whether other grains contained organic phosphoric acids identical with those previously described or if compounds of different composition were present.

In the present investigation the substance was isolated as a barium salt from 0.2 per ct. hydrochloric acid extract of oats by precipitating with barium chloride. The substance was then repeatedly precipitated from dilute hydrochloric acid alternately with alcohol and with pure recrystallized barium hydroxide (Kahlbaum) until all bases other than barium were removed and until all the inorganic phosphate was eliminated.

Several preparations were made from different lots of oats. The substances showed absolutely no tendency to crystallize and they were all obtained as snow-white amorphous powders. On analysis these various preparations gave fairly concordant results but the composition differed considerably from that required for salts of phytic acid. The preparations were reprecipitated and subjected to various other treatments but were always recovered without showing any great variation in composition and it was therefore thought the substance was homogeneous.

However, it was found finally that these preparations, obtained by direct precipitation, were mixtures of barium salts; probably

[1] Wis. Agr. Exp. Sta., Research Bull. 9, 1910.
[2] Journ. Biol. Chem. 13:311, 1912, and N. Y. Agr. Exp. Sta., Tech. Bull. 25, 1912; and preceding article.
[3] Journ. Biol. Chem. 12:477, 1912, and N. Y. Agr. Exp. Sta., Tech. Bull. 22, 1912.

of two different organic phosphoric acids. Only one, however, has been isolated in pure form. By treating the above mentioned amorphous barium salts with small quantities of cold water it was possible to effect a separation into two preparations having entirely different compositions. After the water-soluble portion had been removed the insoluble substance was found to crystallize readily in the same manner and in the same crystal-form as the barium salt obtained from the acid extracted from cottonseed meal, viz: in round or globular masses of microscopic needles. Repeated recrystallizations did not alter the composition except as to the percentage of barium. When allowed to crystallize from dilute hydrochloric acid containing barium chloride a salt is obtained which contains from 40 to 42 per ct. of barium; when it is brought to crystallize from dilute hydrochloric acid solutions by the addition of alcohol the salt contains about 38 per ct. of barium. So far as one can judge by crystal-form, composition, properties and reactions, the crystalline salts obtained from oats and cottonseed meal are identical.

The water-soluble substance referred to above could be obtained only as a snow-white amorphous powder. In composition it differed entirely from the crystalline product but very slightly from the compound isolated from wheat bran.[4] Owing to the amorphous nature of the substance, however, it is impossible to say at present whether it is a homogeneous body or merely a mixture of various compounds. We hope to study this matter more closely, particularly in comparison with the wheat bran products which we propose to investigate further.

The composition of the crystalline barium salts obtained from oats and cottonseed meal does not agree with the usually accepted formula for phytic acid, viz: $C_6H_{24}O_{27}P_6$. The analytical results of these preparations would indicate that they are salts of an acid of the formula $C_2H_6O_8P_2$ or a multiple of it; probably $C_6H_{18}O_{24}P_6$. Such an acid would be isomeric or identical with inosite hexaphosphate which was suggested by Suzuki and Yoshimura [5] as the formula for phytic acid. We have always found, however, that the phosphorus in the free acid prepared from the above barium salts is always about 1 per ct. lower than this formula requires. It is possible that this low percentage of phosphorus is due to partial hydrolysis in drying — which seems the more likely as the hydrogen is always found somewhat high. When the free acid is dried at a temperature of 100° or higher it turns perfectly black in color; even on drying at 60° or 78° in vacuum the color darkens perceptibly, which would indicate some decomposition. It will be shown later that hydrolysis actually does take place on drying and that a large

[4] *Loc. cit.*
[5] Coll. of Agric. Tokyo, Bull. 7:495.

percentage of the phosphorus in the dried preparations is present as inorganic phosphoric acid.

The analyses of the barium salts on the other hand agree very closely with the formula $C_2H_4O_8P_2Ba$ or $C_6H_{12}O_{24}P_6Ba_3$. It will be shown also that the barium salts suffer but very slight hydrolysis on drying at a temperature of 105°. Evidently, therefore, it is safer to calculate the formula of the free acid from the barium salts rather than from analyses of the free acid itself.

EXPERIMENTAL PART.

ISOLATION OF THE SUBSTANCE.

Whole ground oats, including grain and hull, were digested over night in 0.2 per ct. hydrochloric acid in porcelain percolators covered on the inside with a double layer of cheesecloth. The next day the substance was percolated with the same strength hydrochloric acid until the extract gave no appreciable precipitate with barium chloride. The extract was then filtered through paper and precipitated by adding a concentrated solution of barium chloride in liberal excess. The precipitate, after settling, was filtered on a Buchner funnel and washed in 30 per ct. alcohol. It was then dissolved in sufficient dilute hydrochloric acid, about 1 or 2 per ct., filtered and the filtrate precipitated with barium hydroxide solution. After settling, filtering and thoroughly washing with water the precipitate was again dissolved in the same strength hydrochloric acid as before, filtered and the filtrate precipitated by adding an equal volume of alcohol. After repeating these operations alternately a second time the substance was twice precipitated from dilute hydrochloric acid, same strength as above, with barium hydroxide (Kahlbaum) which had been recrystallized. It was then further precipitated three times from the same strength hydrochloric acid with alcohol. The final precipitate was filtered and washed free of chlorides with dilute alcohol and then in alcohol and ether and dried in vacuum over sulphuric acid.

The crude precipitate obtained by adding barium chloride to the acid extract of oats contains large quantities of impurities, inorganic phosphates, colored substances, etc., which during the above operations are gradually eliminated. The precipitates obtained at first do not dissolve completely in dilute hydrochloric acid. It is therefore necessary to filter such solutions repeatedly in order to free them from suspended insoluble matter. Finally, however, a product is obtained which is readily and completely soluble in the dilute hydrochloric acid, in which it gives a perfectly colorless solution.

When prepared as mentioned above, the dry substance is a snow-white, amorphous powder. It is very readily soluble in dilute hydrochloric and nitric acid, less so in acetic acid. It is soluble to a considerable extent in cold water. On moist litmus paper it shows a strong acid reaction. Heated with hydrochloric acid and phloroglucine, no appreciable color reaction developed. After boiling with dilute sulphuric acid for several minutes, filtering and neutralizing, it did not reduce Fehling's solution. It contained neither nitrogen nor sulphur and gave no reaction for chlorides. Dissolved in dilute nitric acid it gave no reaction with ammonium molybdate even after being kept at a temperature of 65° for some time and standing at room temperature for several days, showing that inorganic phosphates were absent. Bases, other than barium, could not be detected in 0.5 gram of the substance.

Owing to loss in purification the yield is rather unsatisfactory. In one case 13 grams were obtained from 5 kg. of oats; in another case 20 grams were obtained from 10 kg. In all, four preparations were made, which gave a total of about 140 grams of the barium salt.

Much time was expended in an endeavor to obtain the substance in crystalline form but as already mentioned it showed no tendency whatever to crystallize. The amorphous preparations were therefore analyzed after previous drying to constant weight at 105° in vacuum over phosphorus pentoxide. The following results were obtained:

1st preparation: $C = 8.84$; $H = 1.67$; $P = 15.88$; $Ba = 36.72$ per ct.

2d preparation: $C = 8.27$; $H = 1.47$; $P = 16.28$; $Ba = 37.26$ per ct.

3d preparation: $C = 8.37$; $H = 1.60$; $P = 16.48$; $Ba = 36.79$ per ct.

4th preparation: $C = 8.44$; $H = 1.61$; $P = 16.35$; $Ba = 36.61$ per ct.

These results are fairly concordant but the composition differs considerably from that required for tribarium phytate. Calculated for $C_6H_{18}O_{27}P_6Ba_3$: $C = 6.42$; $H = 1.60$; $P = 16.60$; $Ba = 36.78$ per ct.

FURTHER PURIFICATION OF THE BARIUM SALT.

In order to determine whether the composition of the substance would change on further treatment, the following experiment was tried. A portion of the *first preparation* was used. The barium was precipitated with slight excess of sulphuric acid, the barium sulphate filtered off and the filtrate precipitated with excess of copper acetate. The copper salt was filtered and thoroughly washed in water and then suspended in water and decomposed with hydrogen sulphide. After removing the copper sulphide, the filtrate was boiled to expel hydrogen sulphide and then precipitated with a solution of recrystallized barium hydroxide. Dilute hydrochloric acid was then added until the precipitate was just dissolved and. the solution precipitated by adding an equal volume of alcohol. The precipitate was filtered, washed in dilute alcohol and then dissolved in 0.5 per ct. hydrochloric acid and reprecipitated with alcohol. The precipitate was then filtered, washed in dilute alcohol, alcohol and ether, and dried in vacuum over sulphuric acid. It was analyzed after drying at 105° in vacuum over phosphorus pentoxide.

Found: $C = 8.23$; $H = 1.56$; $P = 16.19$; $Ba = 37.46$ per ct.

In another case 6 grams of the *second preparation* were heated in a sealed tube with 45 c.c. of $^{1.5}/_N$ sulphuric acid in the steam-bath for twenty-four hours and then allowed to stand for four days at room temperature. After isolating in the same manner as above 3.7 grams were recovered. It was analyzed after drying as above.

Found: $C = 8.42$; $H = 1.66$; $P = 16.17$; $Ba = 37.08$ per ct.

These treatments apparently caused no change in composition.

The free acid was then prepared and analyzed. From the *first preparation*, the acid was prepared in the usual way — i. e. the barium was precipitated with slight excess of sulphuric acid, filtered, and the filtrate precipitated with copper acetate. The copper salt was filtered, washed and decomposed with hydrogen sulphide, filtered and evaporated in vacuum at a temperature of 40°–45° and finally dried in vacuum over sulphuric acid. It was thus obtained as a thick, practically colorless syrup. For analysis it was dried to constant weight over boiling chloroform in vacuum over phosphorus pentoxide. The color turned very slightly dark on drying in this way.

Found: $C = 13.24$; $H = 3.26$; $P = 25.50$ per ct.

The acid prepared from the repurified barium salt gave the following result on analysis after drying as above.

Found: $C = 13.17$; $H = 3.39$; $P = 25.48$ per ct.

The composition of the acid agrees with that required for the above barium salts and one might suppose from the close agreement of analytical results that the substance was homogeneous.

It was found, however, that after the barium salt had been precipitated a great number of times from dilute hydrochloric acid by barium hydroxide and alcohol alternately that the composition did change slightly. The same result was also observed on digesting the barium salt in dilute acetic acid. After treating in the above manner, barium salts of the following composition were obtained:

I: $C = 7.49$; $H = 1.63$; $P = 16.77$; $Ba = 37.89$ per ct.
II: $C = 7.69$; $H = 1.47$; $P = 16.75$; $Ba = 37.72$ per ct.
III: $C = 7.26$; $H = 1.75$; $P = 16.45$; $Ba = 36.40$ per ct.

These salts were united and dissolved in the least possible amount of 0.5 per ct. hydrochloric acid and alcohol added to the solution until a faint permanent turbidity remained — which was just cleared up by the addition of a few drops of dilute hydrochloric acid. The solution was then allowed to stand at room temperature for about two days. There separated slowly a heavy white crust on the bottom of the flask. Under the microscope this showed no definite crystalline structure. The substance was filtered off, washed thoroughly in water, alcohol and ether and dried in the air. It was free from chlorides and gave no reaction with ammonium molybdate. For analysis it was dried at 105° in vacuum over phosphorus pentoxide.

Found: $C = 6.07$; $H = 1.35$; $P = 16.77$; $Ba = 40.00$; $H_2O = 10.08$ per ct.

The carbon found is undoubtedly somewhat too low, as this heavy compact substance burned with extreme difficulty.

The filtrate from the above was precipitated with alcohol, the precipitate filtered, washed and dried in vacuum over sulphuric acid. The following result was obtained on analysis:

Found: $C = 7.63$; $H = 1.57$; $P = 16.53$; $Ba = 36.92$ per ct.

The heavy crust-like substance was recrystallized as follows: It was dissolved in dilute hydrochloric acid, filtered and about an equal volume of alcohol added and the mixture allowed to stand over night. The precipitate which was amorphous at first had then changed into a crystalline form. Under the microscope it appeared as very small globules consisting of microscopic needles. It was filtered, washed free of chlorides in dilute alcohol, alcohol and ether and dried in vacuum over sulphuric acid. It was a light, voluminous, snow-white crystalline powder. It was analyzed after drying at 105° in vacuum over phosphorus pentoxide.

Found: $C = 6.60$; $H = 1.50$; $P = 17.33$; $Ba = 37.45$; $H_2O = 8.72$ per ct.

The amorphous preparations which remained were united (total weight 52.5 grams). The substance was rubbed up in a mortar with a small quantity of cold water. A considerable portion of the substance dissolved. The insoluble portion was perfectly white and opaque but it soon changed into a semicrystalline form and appeared translucent. After standing for some time it was filtered and washed in water and finally in alcohol and ether and dried in vacuum over sulphuric acid. When dry it was again treated with water in the same way. These operations were repeated three times.

The filtrates and washings from the above were precipitated by the addition of alcohol and these precipitates reserved for examination as will be described later.

The water-insoluble substance was dissolved in the least possible quantity of dilute hydrochloric acid (about 5 per ct. strength), the free acid was then nearly neutralized with barium hydroxide; the solution filtered and alcohol added until a faint, permanent turbidity remained. A concentrated solution of 20 grams of barium chloride was then added and the whole allowed to stand. The substance soon began to crystallize in the same crystal form as the barium salt from cottonseed meal, viz: in globular masses or bundles of fine microscopic needles. After standing over night the crystals were filtered, washed free of chlorides with water and then in alcohol and ether and allowed to dry in the air. A further crop of the same-shaped crystals was obtained from the mother-liquor by carefully adding alcohol and allowing to stand. After filtering, washing and drying these were added to the first crop.

The substance was recrystallized three times in the same way. It was finally obtained as a light, snow-white crystalline powder. It weighed about 27 grams.

This was analyzed after drying at 105° in vacuum over phosphorus pentoxide.

0.2483 gram subst. lost 0.0308 gram H_2O.

0.2175 gram subst. gave 0.0295 gram H_2O and 0.0507 gram CO_2.

0.1456 gram subst. gave 0.0987 gram $BaSO_4$ and 0.0864 gram $Mg_2P_2O_7$.

Found: C $=$ 6.35; H $=$ 1.51; P $=$ 16.54; Ba $=$ 39.89; H_2O $=$ 12.40 per ct.

It was again recrystallized in the same manner and the following result obtained on analysis after drying as before:

Found: C $=$ 6.23; H $=$ 1.27; P $=$ 16.17; Ba $=$ 41.48; H_2O $=$ 12.99 per ct.

The substance was again dissolved in dilute hydrochloric acid, the solution was filtered and then precipitated by the addition of alcohol. The precipitate was amorphous at first but on standing in the mother-liquor over night, it had changed into the usual crystal-form but the globules and crystals were much smaller. After filtering and washing free of chlorides in dilute alcohol, alcohol and ether the substance was dried in vacuum over sulphuric acid. It was analyzed after drying at 105° in vacuum over phosphorus pentoxide.

0.2320 gram subst. gave 0.0325 gram H_2O and 0.0553 gram CO_2.

0.1466 gram subst. gave 0.0947 gram $BaSO_4$ and 0.0894 gram $Mg_2P_2O_7$.

Found: C $=$ 6.50; H $=$ 1.56; P $=$ 17.00; Ba $=$ 38.01 per ct.

As will be noticed from the above analytical results the composition does not change on repeated recrystallizations. We believe therefore that the substance is homogeneous. The variation in the percentage of barium evidently depends upon the formation of more or less acid salts. All the preparations described showed a strong acid reaction on moist litmus paper which indicates free hydrogen ions.

In the estimation of carbon and hydrogen it was found impossible to obtain a white ash by direct combustion even when heated for a long time — the residue in the boat was invariably dark colored, varying from light gray to quite dark. Plimmer and Page [6] also mention the difficulty of completely burning carbon in the presence of phosphoric acid. The crystalline barium salts described in this paper as well as those from cottonseed meal are particularly hard to burn; on the other hand the amorphous salts burn more easily. In the combustions of these salts we have always burned the substance twice; first in the regular manner; the dark-colored ash has then been powdered in an agate mortar and mixed in the boat with chromic acid and burned a second time. In this second combustion there has been observed an increase in weight in the carbon dioxide varying from about 1 to 12 milligrams. Under these conditions it is impossible to say whether a complete combustion has been affected and it is not improbable that a small quantity of carbon has escaped oxidation. We are inclined to believe that the percentage of carbon as found is slightly low. As a check upon the carbon content found in the barium salts we have always prepared and analyzed the free acid in the combustion of which we have never experienced any serious difficulty, the residue in the boat showing no trace of carbon.

PREPARATION OF THE FREE ACID FROM THE PURIFIED CRYSTALLINE BARIUM SALT.

The acid was prepared from 4 grams of the barium salt in the usual way. After drying in vacuum over sulphuric acid at room temperature it formed a practically colorless, thick syrup. Its reactions were identical with those which we reported for the acid from cottonseed meal.[7] For analysis it was dried in vacuum over phosphorus pentoxide at 78°. The preparation darkened perceptibly in color but did not turn black.

0.4936 gram subst. gave 0.1366 gram H_2O and 0.1960 gram CO_2.
0.1645 gram subst. gave 0.1601 gram $Mg_2P_2O_7$.
Found: $C = 10.82$; $H = 3.09$; $P = 27.12$ per ct.

PREPARATION OF INOSITE FROM THE BARIUM SALT.

The amorphous barium salt was used. Of the dry salt, 9.3 grams were heated in a sealed tube with 25 c.c. $^5/_N$ sulphuric acid to 150°–160° for three hours. After cooling, the contents of the tube were

[6] *Biochem. Journ.* **7**:167, 1913.
[7] *Loc. cit.*

very dark in color and some carbonaceous substance had separated showing that considerable decomposition had taken place. The sulphuric and phosphoric acids were precipitated with excess of barium hydroxide, filtered and washed and the filtrate freed from excess of barium with carbon dioxide. The filtrate was evaporated to small bulk and decolorized with animal charcoal and then evaporated to dryness on the water-bath. The residue was taken up in a little hot water, filtered from traces of barium carbonate and the inosite brought to crystallize by the addition of alcohol and ether. It separated in needles free from water of crystallization. After filtering, washing in alcohol and ether and drying in the air it weighed 1.6 grams which represents a yield of about 75 per ct. of the total carbon. For analysis it was recrystallized seven times in the same manner as above and was finally obtained in beautiful colorless needles free from water of crystallization. It melted at 223° (uncorrected) and gave the reaction of Scherer. It did not lose in weight on drying at 105° in vacuum over phosphorus pentoxide.

0.1442 gram subst. gave 0.0878 gram H_2O and 0.2113 gram CO_2.
Found: C = 39.96; H = 6.81 per ct.
For $C_6H_{12}O_6$ calculated: C = 40.00; H = 6.66 per ct.

HYDROLYSIS OF THE ACID WITH WATER ALONE.

The acid which was used had been prepared from the amorphous barium salt. Two grams of the dry preparation were heated with 25 c.c. of water in a sealed tube to 190° for $3\frac{1}{2}$ hours. After cooling there was no pressure on opening the tube. The content was of dark brown color and a considerable quantity of a black, carbonized substance had separated. The whole was diluted with water and filtered and the phosphoric acid was precipitated with barium hydroxide in excess. The precipitate was filtered off and examined to see if any unchanged barium phytate could be isolated from it. Apparently the acid had been completely decomposed during the heating as no trace of barium phytate could be found.

The filtrate from the barium phosphate was freed from excess of barium by carbon dioxide and the filtrate evaporated to dryness on the water-bath. The residue was a sticky, amber-colored syrup. It was taken up in a small amount of water and washed into an Erlenmeyer flask. On the addition of alcohol the solution turned cloudy but it could not be brought to crystallize by repeated scratching with a glass rod. It was allowed to stand over night when an amber-colored syrupy layer had separated on the bottom. The upper portion of the liquid was poured off and mixed with ether. On standing a further quantity of amber-colored syrup had separated. The liquid was decanted and evaporated on the water-bath until a small syrupy residue remained. On scratching with a glass rod a substance began to crystallize in small prisms. The other syrups were made to crystallize in the same manner. They were then extracted several times with small quantities of alcohol. The residues were then dissolved in hot water and crystallized by the addition of alcohol. After recrystallizing three times it was obtained in small colorless needles. It weighed 0.25 grams. It melted at 222° (uncorrected) and gave the Scherer reaction and was therefore undoubtedly inosite. After recrystallizing it again melted at 222°. The crystals did not contain water of crystallization as there was no loss in weight on drying at 105° in vacuum for one hour. The substance was further identified as inosite by the analysis.

0.1019 gram subst. gave 0.0629 gram H_2O and 0.1492 gram CO_2.
Found: C = 39.93; H = 6.90 per ct.
For $C_6H_{12}O_6$, calculated: C = 40.00; H = 6.66 per ct.

The alcoholic washings from above and the mother-liquor on evaporation left a dark colored, non-crystallizable syrup. This syrup strongly reduced Fehling's solution on boiling.

It is noteworthy that the amount of inosite obtained by cleavage with water is much less than when dilute sulphuric acid is used. The amount of inosite isolated above represents only about one-half the quantity obtained when the hydrolysis is effected in the presence of acid. It is possible, however, that in this case the inosite had been less completely isolated since the adhering syrupy substance rendered crystallization more difficult.

EXAMINATION OF THE WATER-SOLUBLE PORTION OF THE AMORPHOUS BARIUM SALT.

As has been already mentioned on page 17 the filtrates containing the water-soluble portion of the amorphous barium salt were precipitated with alcohol. After filtering and washing, the precipitate was dried in vacuum over sulphuric acid. It was again digested in water three times and filtered from small insoluble matter and the filtrates precipitated with alcohol and dried. It was finally obtained as a snow-white amorphous powder. As it was impossible to obtain any crystalline substance from it, it was analyzed after drying at 105° in vacuum over phosphorus pentoxide.

0.3504 gram subst. lost 0.0294 gram H_2O on drying.
0.3210 gram subst. gave 0.0616 gram H_2O and 0.1342 gram CO_2.
0.2415 gram subst. gave 0.1465 gram $BaSO_4$ and 0.1236 gram $Mg_2P_2O_7$.

Found: C = 11.40; H = 2.14; P = 14.26; Ba = 35.69; H_2O = 8.39 per ct.

The substance was free from inorganic phosphate and chlorides, and bases other than barium could not be detected. We hope to investigate this substance further.

CONCERNING PHYTIN IN CORN.

The organic phosphoric acid compound occurring in corn has been particularly studied by Vorbrodt.[1]   In an exhaustive treatise on the subject he reports analytical results obtained from crystalline barium salts which led him to believe that the substance was different from phytin.   The barium salt corresponded to the formula $C_{12}H_{26}O_{46}P_{11}Ba_7$ according to which the formula of the acid would be $C_{12}H_{40}O_{46}P_{11}$.   Vorbrodt also showed that the substance gave inosite and phosphoric acid on cleavage either with dilute sulphuric acid or water alone in neutral solution.

The same subject has also been investigated by Hart and Tottingham.[2]   They report the preparation and analysis of the free acid.   The analytical data agrees very closely with that required for phytic acid and they concluded that corn contains phytin.   They also showed that the acid yields inosite on hydrolysis in a sealed tube in the presence of dilute sulphuric acid.

We have undertaken to reexamine this substance in the hope of identifying it either with phytic acid or the compounds which we have shown to exist in cottonseed meal[3] and oats.[4]

At first we were unable to obtain the barium salt in crystalline form but the amorphous salt gave results on analysis which approximately agreed with the corresponding barium phytate.   The free acid, prepared from this amorphous compound, gave about one per ct. too high carbon and about 0.8 per ct. too high phosphorus.

We finally succeeded, however, in preparing a crystalline barium salt.   It was purified by repeated recrystallizations until the composition remained constant.   The product was free from inorganic phosphate and it did not contain a determinable quantity of bases other than barium.   Judging by crystal-form, composition and properties the substance is identical with those previously isolated from cottonseed meal[5] and oats.[6]

The analytical results obtained from these purified crystalline barium salts do not agree with the formula proposed by Vorbrodt.[7] We find the phosphorus over 1 per ct. higher and the relation between carbon and phosphorus is as 1:1.   The phosphorus content is also considerably higher than that required for a corresponding salt calculated on the usual phytic acid formula.

The barium salt analyzed by Vorbrodt had been prepared from the previously isolated acid by partially neutralizing with barium

[1] Anzeiger Akad. Wiss. Krakau, 1910, Series A, p. 484.
[2] Wis. Agr. Exp. Sta., Research Bull. 9, 1910.
[3] *Journ. Biol. Chem.* 13:311,.1912, and N.Y. Agr. Exp. Sta., Tech. Bull. 25, 1912.
[4] See preceding article.
[5] *Loc. cit.*
[6] *Loc. cit.*
[7] *Loc. cit.*

hydroxide and concentrating in vacuum. The crystalline salt which then separated was washed, dried and analyzed. Apparently no attempt had been made to recrystallize it and it is probable that the substance had contained small quantities of impurities which might be sufficient to account for the difference in analytical results between his product and the repeatedly recrystallized salts which we have analyzed.

The composition of the purified salts described in this paper agrees more closely with salts of inosite hexaphosphate than with salts calculated on the basis of the usual phytic acid formula.

EXPERIMENTAL PART.

ISOLATION OF THE SUBSTANCE FROM CORN.

The corn used in these experiments was the ordinary corn meal used as cattle feed at this station. Ground corn meal, 3500 grams, was digested in 7 liters of 0.2 per ct. hydrochloric acid over night. It was then strained and filtered and the clear amber-colored filtrate precipitated by adding about $1\frac{1}{2}$ volumes of alcohol. After settling, the precipitate was filtered and washed in dilute alcohol. The precipitate was then dissolved in a small amount of 0.5 per ct. hydrochloric acid and filtered from insoluble matter. This acid solution gave only a very slight precipitate on the addition of alcohol. The substance was therefore transformed into a barium salt by precipitating with barium hydroxide to slight alkaline reaction. After heating on the water-bath for some time the precipitate was filtered and washed in water. It was again dissolved in 0.5 per ct. hydrochloric acid, filtered and reprecipitated with barium hydroxide. After standing over night the precipitate was filtered and washed thoroughly in water. The substance was again dissolved in 0.5 per ct. hydrochloric acid, filtered and then precipitated by the addition of an equal volume of alcohol. The precipitate after settling was filtered and washed in dilute alcohol. The substance was then precipitated three times more in the same manner and after finally filtering, washing in dilute alcohol, alcohol and ether, it was dried in vacuum over sulphuric acid. A white amorphous powder was obtained which weighed 11.8 grams. The substance gave no reaction for chlorides. The dilute nitric acid solution gave no reaction with ammonium molybdate after warming for some time.

The following results were obtained on analysis after drying at 105° in vacuum over phosphorus pentoxide to constant weight.

Found: $C = 7.25$; $H = 1.51$; $P = 16.65$; $Ba = 37.11$ per ct.

The carbon is somewhat high, otherwise the result agrees with the calculated percentages for tribarium phytate, $C_6H_{13}O_{27}P_6Ba_3$.

Calculated: $C = 6.42$; $H = 1.60$; $P = 16.60$; $Ba = 36.78$ per ct.

PREPARATION OF THE FREE ACID FROM THE ABOVE AMORPHOUS BARIUM SALT.

The acid was prepared from 3 grams of the barium salt in the usual way, i. e. the barium was precipitated with slight excess of sulphuric acid, filtered and the filtrate precipitated with copper acetate. The copper precipitate was filtered and washed thoroughly in water, suspended in water and decomposed with hydrogen sulphide, filtered and the filtrate evaporated in vacuum at a temperature of 40° to 45° to a syrupy consistency and finally dried in vacuum over sulphuric acid. The product was a thick, faintly amber-colored syrup. For analysis it was dried at 105° in vacuum over phosphorus pentoxide. It turned very dark in color.

Found: $C = 11.09$; $H = 3.04$; $P = 26.85$ per ct.

Both carbon and phosphorus are higher than required for phytic acid, $C_6H_{24}O_{27}P_6$.

Calculated: $C = 10.08$; $H = 3.36$; $P = 26.05$ per ct.

PREPARATION OF THE SUBSTANCE FROM CORN AS A CRYSTALLINE BARIUM SALT.

A larger quantity of corn meal was extracted with 0.2 per ct. hydrochloric acid, the extract filtered and precipitated by adding a concentrated solution of barium chloride. The precipitate was then purified and crystallized in the manner described for cotton-seed meal in the preceding article. After the substance had been separated from dilute hydrochloric acid solutions twelve times (eleven times in crystalline form) it was obtained as a beautiful snow-white, bulky crystalline powder which weighed 49 grams. The crystal form was identical with that of the barium salts from cottonseed meal and oats, i. e. globular masses of microscopic needles. The substance was free from chlorides and inorganic phosphate and we were unable to detect any metals other than barium.

For analysis it was dried at 105° in vacuum over phosphorus pentoxide.

0.4101 gram subst. gave 0.0607 gram H_2O and 0.0985 gram CO_2.
0.1477 gram subst. gave 0.1035 gram $BaSO_4$ and 0.0860 gram $Mg_2P_2O_7$.
Found: $C = 6.55$; $H = 1.65$; $P = 16.23$; $Ba = 41.23$ per ct.

A portion of this salt was recrystallized as follows: 5 grams were dissolved in a small quantity of 3 per ct. hydrochloric acid; barium hydroxide was carefully added until a slight permanent precipitate remained; the solution was filtered and a concentrated solution of 2 grams barium chloride added. The perfectly clear solution was allowed to stand at room temperature for about 2 days when the substance separated slowly in the usual form. It was filtered, washed free of chlorides with water and then in alcohol and ether and dried in the air; yield 4.5 grams. The substance gave no reaction with ammonium molybdate.

It was analyzed after drying at 105° in vacuum over phosphorus pentoxide.

0.4770 gram subst. lost 0.0561 gram H_2O.
0.1767 gram subst. lost 0.0209 gram H_2O.
0.4209 gram subst. gave 0.0491 gram H_2O and 0.0931 gram CO_2.
0.3616 gram subst. gave 0.0427 gram H_2O and 0.0832 gram CO_2.
0.1553 gram subst. gave 0.1110 gram $BaSO_4$ and 0.0907 gram $Mg_2P_2O_7$.

Found: I. C = 6.03; H = 1.30; P = 16.28; Ba = 42.06 per ct.
II. C = 6.27; H = 1.32;
H_2O = 11.76 and 11.82 per ct.

For heptabarium inosite hexaphosphate $(C_6H_{11}O_{24}P_6)_2Ba_7 = 2267$.

Calculated: C = 6.35; H = 0.97; P = 16.40; Ba = 42.39 per ct
For 16 H_2O calculated: 11.27 per ct.

This recrystallized salt was again recrystallized as follows: it was dissolved in a small quantity of 3 per ct. hydrochloric acid, filtered and diluted with a small quantity of water. Alcohol was then added until a faint permanent cloudiness remained. On standing at room temperature the substance soon began to crystallize in the usual form except that the crystals were much smaller. After standing over night it was filtered, washed in dilute alcohol, alcohol and ether and dried in vacuum over sulphuric acid. The product was a snow-white, fine crystalline powder. It gave no reaction for chlorides and none for inorganic phosphate. It was analyzed after drying at 105° in vacuum over phosphorus pentoxide.

0.3719 gram subst. gave 0.0465 gram H_2O and 0.0887 gram CO_2.
0.1780 gram subst. gave 0.1187 gram $BaSO_4$ and 0.1091 gram $Mg_2P_2O_7$.
Found: C = 6.50; H = 1.40; P = 17.08; Ba = 39.14 per ct.

For tribarium inosite hexaphosphate, $C_6H_{12}O_{24}P_6Ba_3 = 1066$.

Calculated: C = 6.75; H = 1.12; P = 17.44; Ba = 38.65 per ct.

PREPARATION OF THE FREE ACID.

The acid was prepared in the usual way from about 4 grams of the crystalline barium salt. After drying in vacuum over sulphuric acid it was obtained as a thick, very faintly amber-colored syrup. In appearance and reactions it corresponded exactly with the acids prepared from the crystalline barium salts which have been described previously.

A portion of the above dried acid dissolved in water and acidified with nitric acid gave no precipitate after warming for some time with ammonium molybdate solution.

For analysis the preparation was dried first for ten days in vacuum over sulphuric acid at room temperature and then at 78° in vacuum over phosphorus pentoxide to constant weight. The color did not change by drying in vacuum at room temperature but at 78° the color darkened somewhat.

0.3405 gram subst. gave 0.0919 gram H_2O and 0.1356 gram CO_2.
0.1796 gram subst. gave 0.1754 gram $Mg_2P_2O_7$.
Found: C = 10.86; H = 3.02; P = 27.22 per ct.
For inosite hexaphosphate, $C_6H_{18}O_{24}P_6 = 660$.
Calculated: C = 10.90; H = 2.72; P = 28.18 per ct.
For phytic acid, $C_6H_{24}O_{27}P_6 = 714$.
Calculated: C = 10.08; H = 3.36; P = 26.05 per ct.

CONCERNING THE COMPOSITION OF BARIUM PHYTATE
AND PHYTIC ACID FROM COMMERCIAL PHYTIN AND
A STUDY OF THE PROPERTIES OF PHYTIC ACID AND
ITS DECOMPOSITION PRODUCTS.

In previous reports from this laboratory it has been shown that
the organic phosphoric acids existing in cottonseed meal,[1] oats[2]
and corn[3] yield identical crystalline barium salts which differ in
composition from the corresponding, so-called, barium phytates.
The free organic phosphoric acids, isolated from these crystalline
barium salts, although identical so far as analyses are concerned,
differ in composition from phytic acid. These crystalline salts
had all been carefully purified by repeated recrystallizations and
it appears, therefore, reasonable to believe that they were purer
than any previously described barium phytate. In so far as one
may judge by crystal form, composition and reactions the above
compounds are identical and are salts of an acid of the formula,
$C_2H_6O_8P_2$ or $C_6H_{18}O_{24}P_6$. If the latter be the correct formula, which
appears probable, then it differs from the phytic acid formula of
Neuberg,[4] $C_6H_{24}O_{27}P_6$, by three molecules of water, which is also
the difference between phytic acid and inosite hexaphosphate.

Previously we have reported[5] numerous salts of phytic acid
prepared from commercial phytin. These salts, however, were
mostly amorphous and particularly the barium salts, with one
exception, were not obtained in crystalline form. These amorphous
compounds gave results on analyses which corresponded closely
with percentages calculated on the basis of the usual formula for
phytic acid, viz: $C_6H_{24}O_{27}P_6$. In dealing with amorphous substances, however, some doubt may be felt as to their being homogeneous products.

Since the crystalline salts mentioned above differ in composition
from compounds calculated on the usual formula for phytic acid
we are forced to the conclusion, either that the organic phosphoric
acid existing in cottonseed meal, oats and corn is different and distinct from phytic acid or else that the formula of phytic acid itself
is wrong, having possibly been based upon analytical data of somewhat impure preparations.

It seemed of importance to determine whether any real difference
exists between the barium salts of phytic acid prepared from commercial phytin and the crystalline salts obtained from cottonseed

[1] *Journ. Biol. Chem.* 13:311, 1912, and N.Y. Agr. Exp. Sta., Tech. Bull. 25, 1912.
and also preceding article.
[2] See preceding article.
[3] See preceding article.
[4] *Biochem. Zeitschr.* 9:551, 557, 1908.
[5] *Jour. Biol. Chem.* 11:471, 1912, and 12:97, 1912, and N.Y. Agr. Exp. Sta., Tech.
Bull. 19 and 21, 1912.

meal, oats and corn. We have, therefore, re-examined the commercial phytin using some of the same preparation as before.

After carefully purifying the barium salt of the substance we found that it crystallized very readily and no difference could be observed either in crystal form, composition or reactions, of the salts prepared in this way, from the crystalline salts previously referred to. All of these compounds are therefore identical and the analytical data indicate that they are salts of the acid $C_2H_6O_8P_2$ or $C_6H_{18}O_{24}P_6$.

The composition, as determined by analysis, of the free acid prepared from the crystalline barium phytate also agrees more closely with the above formulas than with the usual formula of phytic acid, $C_6H_{24}O_{27}P_6$. The phosphorus was found too low in this case as well as in the acids previously described. This, however, is undoubtedly due to the fact that the acid is largely hydrolyzed on drying.

It appears very probable then that the organic phosphoric acid described above and known as phytic acid is either inosite hexaphosphate, $C_6H_{18}O_{24}P_6$, or else an isomer of the same. We have, however, no direct information concerning the molecular magnitude of the acid.

We have endeavored to prepare a neutral ester of the acid with which molecular weight determinations might be made, but so far these attempts have failed. By acting on the silver salt of the acid, suspended in absolute methyl alcohol, with methyl iodide, an ester is formed but it apparently suffers partial decomposition in drying. Moreover we have been unable to prepare a neutral silver salt. Only acid silver salts have been obtained even from solutions of phytic acid neutralized with ammonia. From such salts, naturally, only acid esters could be obtained.

As has been pointed out by Starkenstein,[6] which observation we have confirmed,[7] apparently only one-half of the acid hydroxyls of phytic acid are particularly reactive. Some of the acid hydroxyls appear to be very weak. It is no doubt due to this fact that from a neutral solution of ammonium phytate only acid silver salts are precipitated with silver nitrate.

Starkenstein[8] reported that the commercial phytin which he had examined contained relatively large quantities of inorganic phosphate and that it also contained free inosite and he concluded that the substance undergoes spontaneous decomposition. He also found that after drying the preparation at 100° the greater portion of the phosphorus was present as inorganic phosphate.

These results, so far as the formation of inosite on mere drying is concerned, we could not confirm[9] in the phytin preparations

[6] *Biochem. Zeitschr.* 30:65, 1910.
[7] *Journ. Biol. Chem.* 11:475, 1912, and N.Y. Agr. Exp. Sta. Tech. Bull. 19, 1912.
[8] *Loc. cit.*, pp. 59 and 60.
[9] *Journ. Biol. Chem.* 11:473, 1912, and N.Y. Agr. Exp. Sta. Tech. Bull. 19, 1912.

which we had on hand. From a sample of commercial phytin which had been in our laboratory for several years we could not isolate a trace of inosite either before or after drying at 115°. At that time we made no effort to determine the increase in inorganic phosphate on drying at 100° or higher.

Observations made since then, however, have shown without any doubt that phytin undergoes spontaneous decomposition when kept under ordinary conditions at room temperature. Both the salts and the free acid decompose slowly, with liberation of inorganic phosphate. The free acid decomposes much faster than the salts. We have also found that a very perceptible increase in inorganic phosphate occurs on drying at 105° in vacuum. In this case also the free acid decomposes to a greater extent than the salts.

Although notable quantities of inorganic phosphoric acid are liberated from phytic acid and its salts under the above conditions, we have again been unable to demonstrate the presence of inosite as one of the spontaneous decomposition products. In this connection we especially examined a specimen of phytic acid which had been kept in the laboratory for about 18 months. The preparation had been kept in a glass-stoppered bottle at ordinary temperature but at no time had it been exposed to direct sunlight. When first prepared the acid was a practically colorless, thick syrup containing about 20 per ct. of water and it gave no reaction with ammonium molybdate. It darkened gradually in color and when examined the color was quite black. Analysis showed that about one-eighth of the total phosphorus was present in the form of inorganic phosphoric acid. A quantity of this preparation corresponding to 10 grams of the dry acid was examined for inosite but no trace of this substance could be found although the preparation should have contained about 0.3 gram of inosite had the organic radical corresponding to the free inorganic phosphoric acid present separated in the form of inosite.

Since the organic part of the phytic acid radical had not separated as inosite under the above conditions of spontaneous decomposition it appeared of interest to determine, if possible, what product or products had been formed and in what manner the decomposition had occurred. While we are unable to answer these questions fully at this time, the results would indicate that, under the above conditions, the phytic acid undergoes only partial decomposition with formation of penta- or tetra-phosphoric acid esters of inosite and free phosphoric acid.

The aqueous solution of the above partially decomposed phytic acid was precipitated with barium hydroxide. The barium precipitate was freed from inorganic phosphate in our usual way, i. e. by precipitating its dilute hydrochloric acid solution with alcohol until the product gave no reaction with the ammonium molybdate reagent. The final product was a white, amorphous powder. We succeeded in separating this substance into two portions, one a crystalline salt showing all the characteristics and composition of unchanged barium phytate and a second amorphous portion which, judging by analysis, was probably a mixture of the barium salts of inosite penta- and tetra-phosphate.

Since no inosite could be isolated from this preparation, although it had been standing for 18 months and had undergone considerable decomposition, it would seem that a very long time would be required for complete decomposition, i. e., until free inosite were present. On drying a sample of the same acid at 105° for 48 hours under diminished pressure, however, the decomposition products isolated were found to be quite different. In this case we found that about 75 per ct. of the phosphorus was present as inorganic phosphoric acid and we were unable to isolate any unchanged barium phytate. Apparently all of the phytic acid had been partially decomposed and some of it completely, for we obtained 0.25 gram of inosite from 10 grams of the acid after drying as above. The organic phosphoric acid or acids remaining undecomposed were isolated as barium salts. None of these, however, could be obtained in pure form. But by taking advantage of their varying solubilities in water and mixtures of acidulated water and alcohol we were able to separate it into four fractions. All of these fractions had different composition and it would appear probable that they represent more or less impure mixtures of the barium salts, of tetra-, tri-, di-, and mono-phosphoric acid esters of inosite.

When phytic acid has been completely dried at temperatures ranging from 60° to 105° we have noticed that it is not completely soluble in water. Some insoluble substance separates in thin gelatinous plates. We have not been able to obtain a sufficient quantity of this substance for a complete examination. Judging by the analysis of one small sample it is a complex decomposition product of phytic acid, possibly a partially dehydrated tri-phosphoric acid ester of inosite.

EXPERIMENTAL PART.

PREPARATION OF THE CRYSTALLIZED BARIUM PHYTATE.

A sample of the same commercial phytin as formerly examined was transformed into the barium salt as follows: 50 grams of the substance were suspended in about 1500 cubic centimeters of water and dissolved by the careful addition of dilute hydrochloric acid. Barium hydroxide was then added to slight alkaline reaction and the whole allowed to stand over night. The barium hydroxide used was Kahlbaum C. P. which had been recrystallized. The precipitate was filtered and washed thoroughly in water. It was dissolved in the minimum quantity of about 3 per ct. hydrochloric acid, filtered and again precipitated with barium hydroxide. These operations were repeated four times. The substance was then precipitated from the same strength hydrochloric acid with alcohol. After thoroughly washing the precipitate with dilute alcohol it was again dissolved in 3 per ct. hydrochloric acid and precipitated a fifth time with barium hydroxide. The dilute hydrochloric acid solution of the substance was then twice precipitated with alcohol. After finally filtering, the precipitate was washed free of chlorides with dilute alcohol and then washed in alcohol and ether and dried in vacuum over sulphuric acid. The substance was then a snow-white amorphous powder. The dry powder was rubbed up in a mortar with a small quantity of cold water. The insoluble portion changed into a semi-crystalline form after a short time. This was filtered and washed thoroughly in water. It was dissolved in the minimum quantity of 3 per ct. hydrochloric acid. A dilute solution of barium hydroxide was then added until a slight permanent precipitate remained which was nearly cleared up by the careful addition of dilute hydrochloric acid. The solution was filtered and allowed to stand over night. The substance soon began to separate in crystalline form. Under the microscope it appeared perfectly homogeneous and the crystal form was identical with that observed with the barium salts from cottonseed meal, oats and corn, i. e. the substance crystallized in globular masses of microscopic needles.

The substance was filtered off and washed free of chlorides with water and then in alcohol and ether and dried in vacuum over sulphuric acid.

To the mother-liquor a concentrated solution of 15 grams of barium chloride was added and allowed to stand for another 24 hours. A further quantity of the same-shaped crystals had then separated which were filtered, washed and dried as above.

The two crystalline portions were united and recrystallized in the same manner and again dried in vacuum over sulphuric acid. It was then dissolved in the same strength hydrochloric acid and precipitated by adding an equal volume of alcohol. The precipitate was amorphous at first but after standing a few hours it had changed into the crystalline form — identical with the above but the crystals were much smaller. After standing over night it was filtered, washed free of chlorides with dilute alcohol and then in alcohol and ether and dried in vacuum over sulphuric acid.

The product was a snow-white, light, bulky crystalline powder. It weighed 24 grams. It was free from chlorides and inorganic phosphate. In 0.5 gram of the substance no bases other than barium could be detected.

For analysis it was dried at 105° in vacuum over phosphorus pentoxide.

0.2931 gram subst. gave 0.0365 gram H_2O and 0.0714 gram CO_2.
0.1743 gram subst. gave 0.1144 gram $BaSO_4$ and 0.1073 gram $Mg_2P_2O_7$.
Found: C = 6.64; H = 1.39; P = 17.15; Ba = 38.62 per ct.
For tribarium inosite hexaphosphate $C_6H_{12}O_{24}P_6Ba_3$ = 1066.
Calculated: C = 6.75; H = 1.12; P = 17.44; Ba = 38.65 per ct.

The above salt was recrystallized as follows: 5 grams were dissolved in the least possible quantity of about 3 per ct. hydrochloric acid and the free acid nearly neutralized by the careful addition of barium hydroxide until a faint permanent precipitate remained. The solution was then allowed to stand over night. The substance had then separated in the same crystal form as before. It was filtered, washed free of chlorides with water and then in alcohol and ether and allowed to dry in the air.
The product was a heavy, crystalline, snow-white powder. Its dilute nitric acid solution gave no reaction with ammonium molybdate.
For analysis it was dried in vacuum over phosphorus pentoxide at 105°.

0.5772 gram subst. gave 0.0642 gram H_2O.
0.2124 gram subst. gave 0.0234 gram H_2O.
0.5130 gram subst. gave 0.0547 gram H_2O and 0.1180 gram CO_2.
0.1887 gram subst. gave 0.1348 gram $BaSO_4$ and 0.1098 gram $Mg_2P_2O_7$.
Found: C = 6.27; H = 1.19; P = 16.22; Ba = 42.03; H_2O = 11.12 and 11.01 per ct.

For heptabarium inosite hexaphosphate:

$(C_6H_{11}O_{24}P_6)_2Ba_7$ or $C_{12}H_{22}O_{48}P_{12}Ba_7$ = 2267.
Calculated: C = 6.35; H = 0.97; P = 16.40; Ba = 42.39 per ct.
For $16H_2O$ calculated = 11.27 per ct.

PREPARATION OF THE FREE ACID.

The acid was prepared from 5 grams of the first crystalline barium salt in the usual way, i. e., the substance was suspended in water and the barium removed by a slight excess of dilute sulphuric acid, filtered and the filtrate precipitated with excess of copper acetate. The copper precipitate was filtered, washed thoroughly in water, suspended in water and the copper removed with hydrogen sulphide. The filtrate was then evaporated to small bulk in vacuum at a temperature of 40° to 45° and finally dried in vacuum over sulphuric acid. There remained a practically colorless, thick syrup. The dilute aqueous solution of the acid gave no reaction for inorganic phosphoric acid with ammonium molybdate; the concentrated aqueous solution gave a pure white, crystalline precipitate with ammonium molybdate which, standing at room temperature, remained unchanged for many months but which quickly turned yellowish in color on heating. The reactions of the acid with bases were identical with those previously reported.

For analysis it was dried in vacuum over phosphorus pentoxide to constant weight at the temperature of boiling alcohol. In drying at this temperature the color darkened somewhat.

0.3830 gram subst. gave 0.1106 gram H_2O and 0.1517 gram CO_2.
0.1839 gram subst. gave 0.1802 gram $Mg_2P_2O_7$.
Found: C = 10.80; H = 3.23; P = 27.31 per ct.
For inosite hexaphosphate $C_6H_{18}O_{24}P_6 = 660$.
Calculated: C = 10.90; H = 2.72; P = 28.18 per ct.
For phytic acid according to Neuberg, $C_6H_{24}O_{27}P_6 = 714$.
Calculated: C = 10.08; H = 3.36; P = 26.05 per ct.

CONCERNING SOME CHEMICAL PROPERTIES OF PHYTIC ACID.

SPONTANEOUS LIBERATION OF INORGANIC PHOSPHORIC ACID AT ORDINARY TEMPERATURE AND ON DRYING.

Freshly prepared phytic acid is a practically colorless syrup — especially when, in the concentration of its aqueous solution, the temperature is not allowed to rise above 50°. When the acid has been prepared from a pure salt, free from inorganic phosphate, the free acid does not give any reaction with ammonium molybdate for inorganic phosphoric acid. Whenever such colorless specimens of phytic acid are preserved for any length of time the color always darkens. The change in color is more rapid when the concentrated aqueous solution is allowed to stand exposed to the air or preserved in a well stoppered bottle, than when the acid is kept in the desiccator; but even under the latter condition the color gradually deepens to light yellow, deep yellow, light brown and finally, after several months, to dark brown or black. When the acid is dried for analysis either in vacuum or in an air bath the color darkens very materially in a short time, especially when dried at 100° or higher. When dried at a temperature of 60° or 78° in vacuum the color darkens somewhat, but very slightly in comparison with that produced at higher temperatures.

Patten and Hart [10] asserted that the acid turned dark in color on drying at 110° without undergoing any decomposition. As mentioned by Vorbrodt [11] the grounds for this statement are not

[10] *Am. Chem. Journ.* 31:570, 1904.
[11] Anzeiger Akad. Wiss. Krakau, 1910, Series A, p. 484.

quite clear. A striking change in color such as phytic acid suffers in drying or on mere keeping either in the desiccator or under ordinary conditions would very likely indicate a more or less serious decomposition.

In order to determine to what extent decomposition occurs it was decided to make a series of inorganic phosphoric acid determinations by the usual molybdate method on phytic acid preparations before and after drying. While absolute accuracy could hardly be expected or claimed for this method, at least comparable results would be obtained when the precipitations were done under similar conditions.

One portion of the acid was dried at 105° in vacuum over phosphorus pentoxide to constant weight. It was then dissolved in water, neutralized with ammonia, acidified with nitric acid, ammonium nitrate added and heated to 65°. Ammonium molybdate was then added and kept at above temperature for 1 hour. The precipitate was then determined as magnesium pyrophosphate in the usual way.

Another portion was treated in the same manner without drying, the amount of moisture found on drying as above being deducted from the weight taken.

The acid analyzed on page 32 was used for the first determinations. The fresh preparation, dried in vacuum over sulphuric acid as described, contained about 15 per ct. of water and it gave no reaction with ammonium molybdate. It was allowed to stand in the laboratory at summer temperature (about 80° or 90° Fahr.) in a loosely covered dish for three or four weeks. The color had then changed to light brown. On drying at 105° in vacuum over phosphorus pentoxide for about 24 hours to constant weight it lost about 22 per ct. of its weight, showing that it had absorbed about 7 per ct. of water during this time. The acid, page 32, contained 27.31 per ct. of phosphorus. The dried preparation gave the following as inorganic phosphate:

0.2508 gram dry subst. gave 0.0696 gram $Mg_2P_2O_7$, equivalent to 7.73 per ct. phosphorus or 28.30 per ct. of the total phosphorus was precipitated as inorganic phosphoric acid.

Before drying:

0.1889 gram (dry subst. calculated) gave 0.0039 gram $Mg_2P_2O_7$, equivalent to 0.57 per ct. of phosphorus or 2.08 per ct. of total phosphorus.

As will be noticed from the above figures, 26.2 per ct. of the total phosphorus had been hydrolyzed by drying at 105° for about 24 hours.

An old sample of phytic acid which had been kept in the laboratory for about 18 months was examined in the same manner. It was practically black in color. It lost about 22 per ct. of its weight on drying as above for about 20 hours. After decomposing by the Neumann method it was found to contain 27.68 per ct. of phosphorus.

The dry preparation gave the following:

0.2348 gram dry subst. gave 0.0733 gram $Mg_2P_2O_7$ equivalent to 8.70 per ct. of phosphorus, or 31.43 per ct. of the total phosphorus was present as inorganic phosphoric acid.

Before drying:

0.2651 gram (dry subst. calculated) gave 0.0295 gram $Mg_2P_2O_7$ equivalent to 3.10 per ct. of phosphorus or 11.19 per ct. of the total phosphorus had been hydrolyzed in about 18 months under ordinary room conditions.

In the above case about 20.2 per ct. of the total phosphorus had been hydrolyzed on drying at 105° for about 20 hours.

A sample of the pure recrystallized barium phytate was examined for inorganic phosphoric acid in the same way. The fresh preparation gave no reaction with ammonium molybdate. After standing in the laboratory for five or six weeks the following results were obtained:

After drying at 105° in vacuum over phosphorus pentoxide 0.2108 gram subst. gave 0.0104 gram $Mg_2P_2O_7$ equivalent to 1.37 per ct. of phosphorus.

Before drying:

0.2060 gram (dry subst. calculated) gave 0.0030 gram $Mg_2P_2O_7$ equivalent to 0.40 per ct. of phosphorus.

By drying at 105° the inorganic phosphorus increased about $2\frac{1}{2}$ times.

A portion of the inorganic phosphoric acid found in the above determinations was probably due to cleavage of the phytic acid by the dilute nitric acid. Such cleavage appears to take place slowly and uniformly as shown by the following experiment: Another portion (0.1876 gram subst.) of the same barium phytate without previous drying gave 0.0022 gram $Mg_2P_2O_7$ after heating one hour with the ammonium molybdate, or 0.32 per ct. inorganic phosphorus; after heating the solution $\frac{1}{2}$ hour more 0.0010 gram $Mg_2P_2O_7$ was obtained; further heating for 1 hour gave 0.0022 gram $Mg_2P_2O_7$ and a fourth hour heating gave 0.0040 gram $Mg_2P_2O_7$. The total inorganic phosphorus obtained after heating $3\frac{1}{2}$ hours as above was 1.39 per ct. The results indicate that the cleavage under these conditions is slow and that it proceeds at a very uniform rate.

EXPERIMENT TO DETERMINE WHETHER INOSITE IS FORMED IN THE SPONTANEOUS DECOMPOSITION OF PHYTIC ACID.

The sample of old phytic acid previously referred to was used. As shown by the analysis on page 34, the preparation contained 3.10 per ct. inorganic phosphorus. Of this acid, 12.8 grams (corresponding to 10 grams of the dry substance) were dissolved in about 500 cubic centimeters of water and barium hydroxide (Kahlbaum, alkali free) added to slight alkaline reaction. The precipitate was filtered and washed several times in water. The barium precipitate was reserved for special examination.

The filtrate was examined for inosite as follows: The excess of barium hydroxide was precipitated with carbon dioxide, filtered and evaporated on the water bath nearly to dryness. The residue was taken up in a few cubic centimeters of hot water, filtered from a small amount of barium carbonate and the filtrate mixed with alcohol and ether and allowed to stand for several days in the ice chest. A trace of a white amorphous precipitate had separated but absolutely no inosite crystals appeared.

In case the organic part of the phytic acid molecule, corresponding to the inorganic phosphoric acid present, had separated as inosite the above quantity, 10 grams, should have contained about 0.3 gram of inosite and such a quantity could not have escaped detection. Since no inosite could be isolated it seems fair to assume that under the above conditions of spontaneous decomposition phytic acid does not decompose into inosite and phosphoric acid but into phosphoric acid + some unknown substance.

EXAMINATION OF THE ABOVE BARIUM PRECIPITATE.

In the hope of throwing some light upon the nature of this unknown substance the barium precipitate obtained on the addition of barium hydroxide was examined as follows: It was rubbed up with about 400 cubic centimeters of 0.5 per ct. hydrochloric acid and brought into solution by the careful addition of dilute hydrochloric acid. After filtering, it was precipitated by adding an equal volume of alcohol. The precipitate was filtered, washed in dilute alcohol, dissolved in 0.5 per ct. hydrochloric acid and reprecipitated by barium hydroxide. The substance was then precipitated twice from 0.5 per ct. hydrochloric acid with alcohol, finally filtered, washed in dilute alcohol, alcohol and ether and dried in vacuum over sulphuric acid. A white amorphous powder was obtained which weighed 14 grams. It was free from chlorides and inorganic phosphate. After drying at 105° in vacuum over phosphorus pentoxide the following results were obtained.

Found: C = 7.83; H = 1.46; P = 16.72; Ba = 36.96 per ct.

The carbon found is much too high for a pure barium phytate.

PREPARATION OF CRYSTALLIZED BARIUM PHYTATE FROM THE ABOVE AMORPHOUS BARIUM SALT.

The substance was rubbed up in a mortar with about 150 cubic centimeters of cold water and allowed to stand for several hours.

The insoluble portion was changed slowly into a semi-crystalline precipitate. It was filtered and washed in water and then recrystallized as follows: It was dissolved in a small quantity of about 3 per ct. hydrochloric acid, the free acid nearly neutralized with barium hydroxide; a concentrated solution of 10 grams of barium chloride was added, the solution filtered and alcohol added gradually with constant shaking until a slight permanent cloudiness was produced. On standing the substance crystallized slowly in the usual crystal form, i. e., in globular masses of microscopic needles. After two days the crystals were filtered off, washed free of chlorides in water and then in alcohol and ether and dried in the air. Yield 4.5 grams. The substance gave no reaction with ammonium molybdate.

A further quantity of the same-shaped crystals was obtained from the aqueous solution containing the water-soluble portion of the amorphous salt by adding to it 2.5 grams barium chloride and allowing to stand over night. The balance of the water-soluble portion of the substance was recovered by precipitating with an equal volume of alcohol. The resulting precipitate was filtered, washed free of chlorides with dilute alcohol, alcohol and ether and dried in vacuum over sulphuric acid. Yield 4.1 grams.

These substances were analyzed after drying at 105° in vacuum over phosphorus pentoxide.

The recrystallized salt gave the following results:

Found: $C = 6.28$; $H = 1.28$; $P = 15.93$; $Ba = 42.18$; $H_2O = 11.81$ per ct.

The crystalline salt which separated from the aqueous solution gave

$C = 6.47$; $H = 1.23$; $P = 15.95$; $Ba = 42.77$; $H_2O = 12.62$ per ct.

These substances are therefore nearly pure heptabarium salts of inosite hexaphosphate.

Calculated for $(C_6H_{11}O_{24}P_6)_2Ba_7 = 2267$.
$C = 6.35$; $H = 0.97$; $P = 16.40$; $Ba = 42.39$ per ct.

The water-soluble substance precipitated with alcohol gave the following:

Found: $C = 8.58$; $H = 1.62$; $P = 15.86$; $Ba = 38.28$ per ct.

This substance was again treated with about 100 cubic centimeters of cold water, the insoluble portion filtered off and the filtrate, after adding 1 gram of barium chloride, precipitated with alcohol. After washing in dilute alcohol, alcohol and ether and drying in vacuum over sulphuric acid 1.4 grams of a white amorphous substance was obtained. For analysis it was dried at 105° in vacuum over phosphorus pentoxide.

Found: C = 8.08; H = 1.68; P = 15.64; Ba = 39.75 per ct.

This water-soluble substance apparently represents a mixture of the barium salts of penta- and tetraphosphoric acid esters of inosite.

CONCERNING THE DECOMPOSITION PRODUCTS OF PHYTIC ACID AFTER DRYING AT 105° UNDER REDUCED PRESSURE.

The specimen of old phytic acid previously examined was used; 12.8 grams (corresponding to 10 grams dry acid) was dried at 105° for about 48 hours over sulphuric acid under slightly reduced pressure. It was then dissolved in about 200 c.c. of cold water. The solution was practically black in color and contained particles of carbonized material. It was decolorized by shaking with animal charcoal. The clear, colorless solution was then precipitated with barium hydroxide to slight alkaline reaction, the precipitate filtered and washed in water and reserved for examination. The filtrate and washings were freed from barium with carbon dioxide and evaporated on the water bath to dryness. The residue was taken up in a small amount of hot water and filtered. On adding a little alcohol a heavy voluminous white amorphous precipitate was produced. This was removed from the solution by adding about 3 volumes of alcohol. The precipitate settled, leaving a clear supernatant liquid; adding more alcohol produced no further precipitate. It was then filtered and washed in alcohol and the filtrate reserved

After drying, the above precipitate was obtained as a heavy, white amorphous powder. It was free from inorganic phosphorus but contained barium and after combustion the ash gave a heavy yellow precipitate with ammonium molybdate. This substance was purified as will be described later.

The filtrate from the above precipitate was again evaporated on the water bath nearly to dryness, taken up in hot water, filtered and mixed with alcohol and ether. On scratching with a glass rod a substance began to crystallize in needles. It was allowed to stand in the ice chest over night. The crystals were then filtered, washed in alcohol and ether and dried in the air. Yield 0.25 gram. The substance was recrystallized four times in the same manner and was finally obtained in colorless needles free from water of crystallization. It gave the reaction of Scherer and melted at 222° (uncorrected). It was, therefore, no doubt pure inosite. This was further confirmed by the analysis:

0.1215 gram subst. gave 0.0737 gram H_2O and 0.1780 gram CO_2.
Found: C = 39.95; H = 6.78 per ct.
For $C_6H_{12}O_6 = 180$.
Calculated: C = 40.00; H = 6.66 per ct.

PURIFICATION OF THE BARIUM AND PHOSPHORUS CONTAINING PRECIPI-
TATE REMOVED FROM THE INOSITE SOLUTION WITH ALCOHOL.

The substance mentioned above, precipitated with alcohol, was apparently the barium salt of an organic phosphoric acid but it differed in solubility from any other salt of this nature previously observed. It was very soluble in water and was not precipitated from the aqueous solution by barium hydroxide. The dry substance weighed 1.2 grams. It was dissolved in a small quantity of water, a few drops of dilute hydrochloric acid added and 10 cubic centimeters of $^N/_1$ barium chloride. The solution was heated to boiling and alcohol added until a slight cloudiness was produced. On standing in the cold over night a small amount of a hard crust had separated on the bottom of the flask. This was removed and the solution again heated and more alcohol added when a further quantity separated in the same way. The substance was finally filtered and washed thoroughly in 80 per ct. alcohol, alcohol and ether and dried in the air. Without further purification the substance was analyzed after drying at 105° in vacuum over phosphorus pentoxide.

Found: C = 20.03; H = 3.58; P = 8.53; Ba = 25.55 per ct.

The small quantity precluded recrystallization and we are therefore unable to state whether the substance was pure. The analytical result indicates that it was a barium salt of inosite monophosphate and agrees approximately with this formula:

$C_{18}H_{35}O_{27}P_3Ba_2 = 1051.$
Calculated: C = 20.55; H = 3.33; P = 8.84; Ba = 26.16 per ct.

Such a salt could be represented by the following formula:

$$C_6H_{11}O_6 - P\!\!=\!\!O \begin{matrix} OH \\ O \end{matrix}$$
$$>Ba$$
$$C_6H_{11}O_6 - P\!\!=\!\!O \begin{matrix} O \\ O \end{matrix}$$
$$>Ba$$
$$C_6H_{11}O_6 - P\!\!=\!\!O \begin{matrix} O \\ OH \end{matrix}$$

EXAMINATION OF THE PRECIPITATE PRODUCED WITH BARIUM
HYDROXIDE AFTER DRYING THE ABOVE ACID.

A portion of the barium precipitate was dried in vacuum over sulphuric acid and then examined for total and inorganic phosphorus in the same way as before:

Found: Total phosphorus (by Neumann method) 9.98 per ct.
Found: Inorganic phosphorus 7.46 per ct.

As will be noticed from these figures 74.76 per ct. of the phosphorus was present as inorganic phosphoric acid.

The substance was freed from inorganic phosphate by precipitating four times with alcohol from 0.5 per ct. hydrochloric acid. After finally drying in vacuum over sulphuric acid 3.2 grams of a snow-white, amorphous powder was obtained. The substance was free from chlorides and inorganic phosphate.

It was shaken up with about 75 cubic centimeters of cold water in which the greater portion dissolved; 10 cubic centimeters of $^N/_1$ barium chloride was added and allowed to stand for several hours; the insoluble portion was then filtered off, washed free of chlorides with water and then in alcohol and ether and dried in vacuum over sulphuric acid. It weighed 0.65 gram.

The filtrate from above containing the water-soluble portion of the substance was acidified with a few drops of dilute hydrochloric acid, heated to boiling and alcohol added until a slight permanent cloudiness remained. On standing over night, a portion had separated in the form of a heavy, granular powder. Under the microscope no definite crystal form could be observed but it appeared to consist of transparent globules. It was filtered off, washed free of chlorides in 30 per ct. alcohol, alcohol and ether and dried in the air. Yield 0.67 gram. It was free from inorganic phosphate.

The mother-liquor from above was precipitated with alcohol. After settling, the precipitate was filtered, washed with dilute alcohol, alcohol and ether and dried in vacuum over sulphuric acid. Yield 1.55 grams. The substance was a snow-white, amorphous powder. It was free from chlorides and inorganic phosphate.

These three different portions were analyzed after drying at 105° in vacuum over phosphorus pentoxide.

The water-insoluble portion gave

$C = 9.40$; $H = 1.65$; $P = 13.76$; $Ba = 39.56$ per ct.

Judging by the analysis this substance consists mainly of the barium salt of inosite tetraphosphate.

The granular powder which separated from the hot dilute hydrochloric acid solution and alcohol on cooling gave the following result:

$C = 12.70$; $H = 2.40$; $P = 13.84$; $Ba = 32.29$; $H_2O = 12.77$ per ct.

This substance appears to be mainly the barium salt of inosite triphosphate, although not pure. It was mixed probably with some barium salt of inosite diphosphate.

The water-soluble portion precipitated with alcohol gave:

$C = 14.07$; $H = 2.31$; $P = 12.92$; $Ba = 33.04$ per ct.

Deducting the barium found, allowing for hydrogen and water, and calculating to the free acid these results became:

$C = 20.88; H = 4.11; P = 19.16$ per ct.

This is approximately the composition of inosite diphosphate, $C_6H_{14}O_{12}P_2 = 340$.
Calculated: $C = 21.17; H = 4.11; P = 18.23$ per ct.

That these substances, separated from the partially decomposed phytic acid, are inosite esters of phosphoric acid and not condensation or other decomposition products is evident from the fact that on complete cleavage inosite is obtained. Unfortunately the amount of each of the above substances was too small to permit of examination in this direction except the last one, viz., the water-soluble product precipitated by alcohol and which analyzed for inosite diphosphate. The remainder (0.94 grams dry substance) was hydrolyzed with dilute sulphuric acid in a sealed tube at 150°–160° for about 2½ hours and the inosite isolated in the usual way. The amount of inosite obtained was 0.28 gram or about 88 per ct. of the theory. The substance gave the reaction of Scherer and melted at 222° (uncorrected) which leaves no doubt that it was pure inosite.

It is evident that all of the barium precipitates described above are mixtures. It could hardly be expected that a complete separation into pure chemical compounds of the salts of these inosite esters could be effected by the method used. The analytical results, however, show that it is possible to isolate from partially decomposed phytic acid certain substances approximating in composition various phosphoric acid esters of inosite which on complete cleavage yield inosite just as does phytic acid itself. This fact, we believe, supports the view previously expressed that phytic acid suffers a gradual and partial decomposition, i. e., molecules of phosphoric acid are eliminated one by one. We believe also that these facts taken in connection with the formation of inosite from phytic acid on mere drying at 105° must be considered as a strong support of the theory that phytic acid is inosite hexaphosphate and not some complex compound as previously held.

ATTEMPT TO PREPARE A METHYL ESTER OF PHYTIC ACID.

The silver salt previously described as hepta-silver phytate [12] was used. Of this salt, 5.4 grams were suspended in 100 cubic centimeters of absolute methyl alcohol and 4 grams of methyl iodide (a little over the required amount) were added and the mixture shaken for several hours, the flask being protected from the light. At the end of this time the white silver phytate had changed into the yellow silver iodide.

[12] *Journ. Biol. Chem.* **12** : 107, 1912, and N. Y. Exp. Sta. Tech. Bull. 21, 1912.

The precipitate was filtered off and washed several times in absolute methyl alcohol and the filtrate several times evaporated in vacuum to dryness under addition of methyl alcohol for the removal of the excess of methyl iodide. The residue was dissolved in methyl alcohol and evaporated to dryness in vacuum over sulphuric acid. The substance was then obtained as a light-yellow-colored, thick syrup of faint, aromatic odor. It was strongly acid in reaction and of sharp acid taste. For analysis it was dried in vacuum at 105° over phosphorus pentoxide. It then turned very dark in color.

0.1985 gram subst. gave 0.0608 gram H_2O and 0.1016 gram CO_2.
Found: C = 13.95; H = 3.42 per ct.

This agrees with a dimethyl ester of phytic acid.

For $C_6H_{16}O_{24}P_6(CH_3)_2 = 688$.
Calculated: C = 13.95; H = 3.19 per ct.

THE WATER-INSOLUBLE SUBSTANCE WHICH SEPARATES FROM PHYTIC ACID AFTER DRYING.

As has been mentioned earlier, phytic acid, which has been dried to constant weight in vacuum over phosphorus pentoxide, is not completely soluble in water. We have observed this insoluble substance in many instances after drying phytic acid at 60°, at 78° and at 105°. It always separates, on adding water to the dry substance, in thin gelatinous plates. It appears to be practically insoluble in hot or cold water. Continued boiling in acidulated water is necessary to dissolve it. It is also insoluble in alcohol and ether.

In order to obtain some knowledge of the composition of this insoluble substance 2.7 grams of phytic acid, containing about 12 per ct. of moisture, were dried to constant weight at 105° in vacuum over phosphorus pentoxide. After treating with water the insoluble portion was filtered, washed thoroughly in water and finally in alcohol and ether and dried in vacuum over sulphuric acid. It was then obtained as a dirty gray powder which weighed 0.23 gram. It was non-hygroscopic. For analysis it was dried at 105° in vacuum over phosphorus pentoxide at which no change in color was noticeable. The substance was burned with copper oxide and the phosphorus determined in the ash.

0.2118 gram dry subst. gave 0.0569 gram H_2O and 0.1357 gram CO_2 and 0.1822 gram $Mg_2P_2O_7$.
Found: C = 17.47; H = 3.00; P = 23.98 per ct.

The quantity of the substance obtained was so small that it was only sufficient for one analysis. Of course, we are unable to state whether it was homogeneous or not but the analytical results agree approximately with inosite triphosphate minus one molecule of water. The substance may therefore be a partial pyrophosphoric acid ester of inosite or it may represent some complex decomposition product of phytic acid.

In conclusion we present a summary of the analytical results of the preceding crystalline barium salts in comparison with the calculated percentages required for the usual phytic acid formula and inosite hexaphosphate.

TABLE I.— BARIUM SALTS CRYSTALLIZED FROM DILUTE HYDROCHLORIC ACID BY THE ADDITION OF ALCOHOL.

	Found				Calculated for	
	Cottonseed meal	Oats	Corn	Commercial phytin	Tri-barium inosite hexaphosphate $C_6H_{12}O_{24}$-P_6Ba_3.	Tri-barium phytate $C_6H_{18}O_{27}$-P_6Ba_3.
	Per ct.	Per ct.	Per ct.	Per ct.	Per ct.	Per ct.
C.............	6.61, 6.59	6.50	6.50	6.64	6.75	6.42
H.............	1.34, 1.44	1.56	1.40	1.39	1.12	1.60
P.............	16.91, 17.08	17.00	17.08	17.15	17.44	16.60
Ba.............	39.57, 38.79	38.01	39.14	38.62	38.65	36.78

TABLE II.— BARIUM SALTS CRYSTALLIZED FROM DILUTE HYDROCHLORIC ACID IN THE PRESENCE OF BARIUM CHLORIDE.

	Found				Calculated for	
	Cottonseed meal	Oats	Corn	Commercial phytin	Hepta-barium inosite hexaphosphate $(C_6H_{11}O_{24}$-$P_6)_2Ba_7$.	Hepta-barium phytate $(C_6H_{17}O_{27}$-$P_6)_2Ba_7$.
	Per ct.	Per ct.	Per ct.	Per ct.	Per ct.	Per ct.
C.............	6.29, 6.03	6.23	6.27	6.27	6.35	6.06
H.............	1.11, 1.18	1.27	1.32	1.19	0.97	1.43
P.............	16.54, 15.80	16.17	16.28	16.22	16.40	15.66
Ba.............	42.06, 42.85	41.48	42.06	42.03	42.39	40.46

44

TABLE III.— THE FREE ACIDS PREPARED FROM THE CRYSTALLINE BARIUM SALTS.

	Found				Calculated for	
	Cottonseed meal	Oats	Corn	Commercial phytin	Inosite hexaphosphate $C_6H_{18}O_{24}$-P_6	Phytic acid according to Neuberg $C_6H_{21}O_{27}$-P_6.
	Per ct.	Per ct.	Per ct.	Per ct.	Per ct.	Per ct.
C..............	10.68	10.82	10.86	10.80	10.90	10.08
H..............	3.09	3.09	3.02	3.23	2.72	3.36
P..............	27.66	27.12	27.22	27.31	28.18	26.05

ACKNOWLEDGMENTS.

The author desires to express his appreciation and thanks to Dr. P. A. Levene of the Rockefeller Institute for Medical Research, New York, N. Y., and to Dr. Thomas B. Osborne of the Connecticut Agricultural Experiment Station, New Haven, Conn., for many suggestions which have been of great value in carrying out this work.

TECHNICAL BULLETIN No. 33.   FEBRUARY, 1914.

New York Agricultural Experiment Station.

GENEVA, N. Y.

PREPARATION, COMPOSITION AND PROPERTIES OF CASEINATES OF MAGNESIUM.

LUCIUS L. VAN SLYKE AND ORRIN B. WINTER.

PUBLISHED BY THE DEPARTMENT OF AGRICULTURE.

PREPARATION, COMPOSITION AND PROPERTIES OF CASEINATES OF MAGNESIUM.

L. L. VAN SLYKE AND O. B. WINTER.

SUMMARY.

1. *Preparation of solution of casein in magnesium hydroxide.*— In preparing magnesium caseinates, the solution of casein in magnesium hydroxide is effected by suspending pure casein in water with an excess of finely-divided magnesium oxide, allowing the mixture to stand several days with occasional agitation.

2. *Preparation and composition of basic magnesium caseinate.*— The magnesium hydroxide solution of casein is made neutral to phenolphthalein with HCl and the solution dialyzed and evaporated to dryness. The preparation contains 1.06 per ct. Mg (1.76 MgO), the theoretical composition being 1.09 per ct. Mg (1.81 MgO); or 1 gram of casein combines with 8.7×10^{-4} gram equivalents of Mg (theoretical, 9.0×10^{-4}). The compound is easily soluble in water and in a 5 per ct. solution of NaCl.

3. *Preparation and composition of neutral magnesium caseinate.*— The magnesium hydroxide solution is made neutral to litmus with HCl and the solution dialyzed and the caseinate precipitated with alcohol. The preparation contains 0.71 per ct. Mg (1.18 MgO), the theoretical composition being 0.67 per ct. Mg (1.12 MgO); or 1 gram of casein combines with 5.8×10^{-4} gram equivalents of Mg (theoretical, 5.6×10^{-4}). The compound is easily soluble in water and in a 5 per ct. solution of NaCl.

4. *Preparation and composition of mono-magnesium caseinate.*— A solution of base-free casein in magnesium hydroxide is treated with HCl just to the first point of precipitation and then dialyzed. Alternate addition of acid and dialysis are repeated, until finally the dialyzed solution forms a permanent precipitate on the addition of any acid. To this solution is added one-third of the amount of acid required for complete precipitation of the casein, the solution filtered and dialyzed and divided into two portions. One portion is used for the preparation of mono-magnesium caseinate by incomplete precipitation with HCl. The preparation contains 0.13 per ct. Mg (0.22 MgO), which is the theoretical composition; or 1 gram of casein combines with 1.1×10^{-4} gram equivalents of Mg. This compound is insoluble in water but soluble in 5 per ct. solution of NaCl; at 65° C. it tends to form strings when drawn out.

5. *Preparation and composition of di-magnesium caseinate.*— To the second portion of the solution mentioned in the preceding paragraph acid-free alcohol is added and a precipitate obtained which contained 0.24 per ct. Mg (0.40 MgO), the theoretical composition of di-magnesium caseinate being 0.26 per ct. Mg (0.44 MgO); or 1 gram of casein combines with 2.1 x 10-⁴ gram equivalent of Mg (theoretical 2.25 x 10-⁴). The compound is quite easily soluble in water and in a 5 per ct. solution of NaCl; at 65° C. it is slightly sticky.

6. These four magnesium caseinates correspond to the four calcium caseinates which have been previously prepared, representing octo-, penta-, di- and mono-caseinates of magnesium.

INTRODUCTION.

In Technical Bulletin No. 26 of this Station, Van Slyke and Bosworth report a study of compounds formed by casein with the elements of several alkaline and alkaline-earth bases. In the case of calcium, for illustration, it was found that at least four caseinates could be formed, (1) mono-calcium caseinate, (2) di-calcium caseinate, (3) neutral (or penta-valent) calcium caseinate, neutral to litmus, and (4) basic (or octo-valent) calcium caseinate, neutral to phenolphthalein. It seemed desirable to undertake a similar study to ascertain whether casein forms corresponding compounds with magnesium. The details of this work are given in this bulletin.

DETAILS OF LABORATORY WORK.

SOLUTION OF CASEIN IN MAGNESIUM HYDROXIDE.

In the preparation of caseinates of the alkaline-earth elements, the first step in the process is to obtain a solution of casein in the hydroxide. In the case of magnesium, its hydroxide is only slightly soluble in water and the solution is so dilute as to have very little effect in dissolving casein. It was found, however, that when casein is suspended in water with an excess of finely-divided magnesium oxide and allowed to stand several days with occasional agitation, enough casein is taken into solution to furnish material which can be used in the preparation of magnesium caseinates.

PREPARATION AND COMPOSITION OF BASIC MAGNESIUM CASEINATE.

Base-free casein (prepared in the manner described in Technical Bulletin No. 26, pp. 8–9) was dissolved in a solution of magnesium hydroxide containing an excess of magnesium oxide in suspension. The mixture was filtered and the filtrate was made neutral to phenolphthalein with hydrochloric acid; the end point was satisfactorily determined by adding the acid slowly until a faintly pinkish color remains for several minutes. The solution was then dialyzed to remove the magnesium chloride formed in neutralization. The dialyzed solution was evaporated to dryness and finally dried at 120° C.

Analysis of the material showed it to contain 1.06 per ct. Mg (equal to 1.76 per ct. MgO). Calculated to correspond with basic calcium caseinate (containing 1.78 per ct. Ca or 2.50 per ct. CaO), this compound should contain 1.08 per ct. Mg (equal to 1.80 per ct. MgO). Expressed in another form, our results indicate that 1 gram of casein combines with 8.72 x 10^{-4} gram equivalents of magnesium (theoretical, 9 x 10^{-4}).

This compound is *basic magnesium caseinate*, the casein being octo-valent. The compound is easily soluble in water, its solution being neutral to phenolphthalein. It is also soluble in a 5 per ct. solution of NaCl.

PREPARATION AND COMPOSITION OF NEUTRAL MAGNESIUM CASEINATE.

Base-free casein was dissolved in a solution of magnesium hydroxide containing an excess of magnesium oxide in suspension. The mixture was filtered and the filtrate was made nearly neutral to litmus with N/10 HCl. The solution was dialyzed and then made neutral to litmus with N/10 HCl; the end-point was determined by adding the acid slowly, until both red and blue litmus paper could be left in the solution several minutes without change of color in either. The solution was then dialyzed again and the casein precipitated with acid-free alcohol; the mixture was then filtered, and the precipitate was washed thoroughly and was finally dried at 120° C.

Determination of the magnesium in this preparation gave 0.71 per ct. Mg (equal to 1.18 per ct. MgO). Calculated to correspond with neutral calcium caseinate (containing 1.10 per ct. Ca or 1.55 per ct. CaO), this compound should contain 0.67 per ct. Mg (equal to 1.12 per ct. MgO). Expressed in another form, our results indicate that 1 gram of casein combines with 5.84 x 10^{-4} gram equivalents of magnesium (theoretical 5.625 x 10^{-4}).

This compound is *neutral magnesium caseinate*, the casein being penta-valent. The compound is easily soluble in water and in a 5 per ct. solution of NaCl.

PREPARATION AND COMPOSITION OF DI-MAGNESIUM CASEINATE.

Solution of base-free casein in magnesium hydroxide containing an excess of magnesium oxide is effected in the manner already described. To the filtered solution, N/50 HCl is added until near the point of precipitation, as shown by a preliminary test (Technical Bulletin No. 26, pp. 15–17). The solution is then dialyzed. Alternate addition of acid and dialysis are repeated several times, until the addition of a small amount of acid to a test portion causes precipitation. The amount of acid necessary for complete precipitation is next determined, and about one-third of this amount is added to the solution. The mixture is then filtered to remove any precipitate that is formed and the filtrate is dialyzed to remove magnesium chloride as completely as possible. This solution, containing di-magnesium caseinate, is divided into two portions, one being used for the preparation of di-magnesium caseinate and the other for the preparation of mono-magnesium caseinate.

To one portion of the solution acid-free alcohol is added and a precipitate of di-magnesium caseinate obtained. This is thoroughly washed with acid-free alcohol and ether and dried at 120° C.

Determination of the magnesium in this preparation showed it to contain 0.24 per ct. Mg (equal to 0.40 per ct. MgO). Calculated to correspond with di-calcium caseinate (containing 0.44 per ct. Ca or 0.62 per ct. CaO), this caseinate should contain 0.26 per ct. Mg (equal to 0.44 per ct. MgO). Expressed in another form, our results indicate that 1 gram of casein combined with 2.14×10^{-4} gram equivalents of magnesium (theoretical, 2.25×10^{-4}).

Di-magnesium caseinate is slightly soluble in water; it is soluble in a 5 per ct. solution of NaCl at 65° C. and at this temperature is slightly sticky.

PREPARATION AND COMPOSITION OF MONO-MAGNESIUM CASEINATE.

In preparing mono-magnesium caseinate, the remaining portion of the solution of di-magnesium caseinate was treated with enough acid to precipitate three-fourths of the casein, the acid being added very slowly and with constant, vigorous agitation. The solution was filtered and the precipitated caseinate washed with water, alcohol and ether, after which it was dried for three days in vacuo over sulphuric acid.

Determination of magnesium gave 0.13 per ct. Mg (equal to 0.22 per ct. MgO). Calculated to correspond with mono-calcium caseinate (0.22 per ct. Ca or 0.31 per ct. CaO), this compound should contain 0.13 per ct. Mg (equal to 0.22 per ct. MgO). Expressed in another form, our results show that 1 gram of casein combines with 1.125×10^{-4} gram equivalents of magnesium, which agrees with the theoretical value.

Mono-magnesium caseinate is insoluble in water, but soluble in 5 per ct. NaCl solution; at 65° C. it shows a tendency to form strings when drawn out.

VALENCY OF CASEIN.

In Technical Bulletin No. 26, Van Slyke and Bosworth have shown from their work with calcium caseinates the combining power of casein in the different compounds prepared and studied by them. It is interesting to compare their results with those obtained with magnesium. In the following table we arrange the results expressed, for the purpose of more direct comparison, in the form of gram equivalents of element per gram of casein.

TABLE I.— VALENCY OF CASEIN AS SHOWN IN MAGNESIUM CASEINATES.

Different caseinates.	Valencies satisfied in each caseinate.	Gram equivalents of Ca x 10-4 per gram of casein.	Gram equivalents of Mg x 10-4 per gram of casein.	Gram equivalents of element x 10-4 per gram of casein corresponding exactly to valencies given in second column.
Mono-.........	1	1.11	1.125	1.125
Di-..........	2	2.21	2.14	2.250
Neutral......	5	5.36	5.84	5.625
Basic........	8	9.00	8.72	9.000

The agreement between the analytical results obtained (third and fourth columns) and those called for (fifth column) by the valencies of casein satisfied in each compound (second column) is marked; also the close agreement between the results obtained with calcium (third column) and those given by magnesium (fourth column) is satisfactory.

TECHNICAL BULLETIN No. 34.   MAY, 1914

New York Agricultural Experiment Station.

GENEVA, N. Y.

I. WHY SODIUM CITRATE PREVENTS CURDLING OF MILK BY RENNIN.

ALFRED W. BOSWORTH AND LUCIUS L. VAN SLYKE.

II. THE USE OF SODIUM CITRATE FOR THE DETERMINATION OF REVERTED PHOSPHORIC ACID.

ALFRED W. BOSWORTH.

PUBLISHED BY THE DEPARTMENT OF AGRICULTURE.

I. WHY SODIUM CITRATE PREVENTS CURDLING OF MILK BY RENNIN.*

ALFRED W. BOSWORTH AND LUCIUS L. VAN SLYKE.

SUMMARY.

1. The addition of sodium citrate to milk in infant feeding is a frequent practice in cases in which the use of normal milk results in the formation of large lumps of tough indigestible curd in the stomach. The favorable results attending such use of sodium citrate have never been explained on the basis of actual investigation.

2. Work previously done by the authors suggested a chemical explanation of the observed facts and led them to test the matter by an experimental study of the action of sodium citrate on milk.

3. The addition of sodium citrate to normal milk increases the amount of soluble calcium in the milk, this increase resulting from a reaction between the calcium caseinate of the milk and sodium citrate, by which is formed sodium caseinate (or calcium-sodium caseinate) and calcium citrate. The reaction is reversible.

4. The curdling of milk by rennin is delayed by the presence of sodium citrate; when there is added 0.400 gm. of sodium citrate per 100 c.c. of milk (equal to 1.7 grains per ounce), no curdling takes place.

5. The curd produced by rennin in the presence of small amounts of sodium citrate (0.050 to 0.350 gm. per 100 c.c. or 0.20 to 1.5 grains per ounce) increases in softness of consistency as the amount of sodium citrate in the milk increases.

6. The results of our work indicate that at the point at which rennin fails to curdle milk we have in place of the calcium caseinate of normal milk a double salt, calcium-sodium caseinate; this double salt, when rennin is added, is changed to a calcium-sodium paracaseinate which, owing to the presence of the sodium, is not curdled.

7. The practice of adding sodium citrate to milk at the rate of 1 to 2 grains of citrate per ounce of milk appears to have a satisfactory chemical basis in the reaction between the sodium citrate and the calcium caseinate of the milk. The amount added is governed by the object in view, viz., whether it is desired to prevent curdling or only modify the character of the curd in respect to softness.

---

* Published also in the *Am. Jour. Diseases of Children,* 7 : 298–304.

INTRODUCTION.

The practice of adding sodium citrate to milk used as infant food has been common for many years. It has found application especially in the treatment of certain types of "feeding-cases" in which untreated milk, after entering the stomach, forms abnormally large chunks of tough curd, shown by Talbot [1] to consist of casein. These lumps of curd may pass practically unchanged through the entire intestinal canal, causing mechanical irritation, which often results in serious interference with the process of normal digestion. Empirical practice has shown that this abnormal curdling of milk may, to some extent, be modified or controlled by the addition of sodium citrate at the rate of 1 or 2 grains per ounce of milk. While various suggestions have been offered to explain the results observed, these have been based so little on demonstrated chemical facts as to partake largely of the nature of guesswork.

In our work [2] on the compounds of casein and paracasein we obtained certain results which appeared to suggest a simple and satisfactory explanation of the marked effect produced by the addition of sodium citrate to milk. Work has been done to test the application of the suggested explanation and the results are presented in this paper.

THEORETICAL CONSIDERATIONS.

In order that the details of our investigation may be more readily understood we will call attention to certain fundamental facts which have been brought out in our former work before we give the details of our present investigation. In the work to which reference is made in the paragraph preceding, the following points may be regarded as being established as far as the data now at hand enable us to reach any conclusions:

1. Casein is a protein showing the characteristic property of an acid in that it combines with metals or bases to form compounds known as caseinates.

2. The molecular weight of casein is 8888, and it can combine with eight equivalents of a monovalent metal or base. For example, the compound of casein containing the largest amount of a monovalent metal like sodium could be represented by the formula Na_8 casein (sodium caseinate); the corresponding calcium compound is Ca_4 casein (calcium caseinate).

---

[1] Talbot, F. B. *Boston Med. and Surg. Jour.*, June 11, 1905, p. 205, and Jan. 7, 1909, p. 13.

[2] Van Slyke and Bosworth. *Jour. Biol. Chem.*, 14 : 206, (1913), and Technical Bull. No. 26, N. Y. Agricultural Experiment Station; also Bosworth: *Jour. Biol. Chem.*, 15 : 231, (1913), and Technical Bull. No. 31, N. Y. Agricultural Experiment Station.

3. Casein is present in milk in combination with calcium as calcium caseinate. It has not been definitely settled yet which particular compound is in milk, but it is probably either tetra-calcium or tri-calcium caseinate.

4. When the calcium caseinate of milk is acted on by rennin, it is changed into another compound called calcium *paracaseinate*. By this action one molecule of calcium caseinate is split into two molecules of calcium paracaseinate. Thus, assuming for the sake of our illustration, that the caseinate present in milk is the tetra-calcium compound, we can represent the change from calcium casinate to calcium paracaseinate in the following manner:

Ca_4 caseinate $= Ca_2$ paracaseinate $+ Ca_2$ paracaseinate.

5. Paracasein, like casein, possesses acid properties, but has a molecular weight of 4444, only one-half that of casein. Paracasein, as an acid, has only one-half the combining power of casein; that is, its highest combining power is equal to four equivalents of a monovalent metal; for example, Na_4 paracasein (sodium paracaseinate), Ca_2 paracasein (calcium paracaseinate).

6. Calcium paracaseinate is *less soluble* than the corresponding calcium caseinate present in milk from which it is formed, and, therefore, it is precipitated as a solid, or, in ordinary language, the milk curdles.

7. If rennin is added to a solution of sodium caseinate, the caseinate is split into two molecules of sodium paracaseinate (for example, Na_8 caseinate $= Na_4$ paracaseinate $+ Na_4$ paracaseinate), but no precipitation or curdling takes place. This is explained by the fact that sodium paracaseinate is very soluble. If, however, to this same solution of sodium paracaseinate we add a small amount of some soluble calcium salt (calcium chlorid, for example), curdling occurs at once, the curd being calcium paracaseinate. This precipitation or curdling is the result of a chemical reaction or double decomposition, which can be illustrated in the following manner:

Sodium paracaseinate (soluble) + calcium chlorid = calcium paracaseinate (insoluble) + sodium chlorid.

This reaction or equilibrium can be made to proceed in either direction at the will of the experimenter; for example, addition of excess of sodium chlorid changes insoluble calcium paracaseinate back into soluble sodium paracaseinate. These facts appeared to us to furnish an explanation of the action of sodium citrate when added to milk, in that there is formed calcium citrate and sodium caseinate, which latter compound is converted by rennin into sodium paracaseinate, a compound so soluble as not to curdle or form a precipitate. This hypothesis furnished the basis of our present investigation.

The facts stated above raise the query whether or not sodium citrate reacts with the calcium caseinate in milk to form sodium caseinate and calcium citrate? If such a reaction takes place, we should be able to determine the amount of calcium thus transferred from the caseinate to the citrate. How this determination can be made we will indicate briefly.

When milk is filtered under pressure through unglazed porous porcelain (in the form of a Chamberland filter), the serum containing the filterable soluble portions passes through the filter, while the insoluble portion, consisting largely of caseinate and insoluble calcium phosphate, remains on the filter. If, therefore, milk treated with sodium citrate is filtered through a Chamberland filter, the amount of calcium in the filtered serum should increase with the amount of citrate added up to a certain point, provided that calcium citrate (or perhaps, a double salt of calcium-sodium citrate) is formed. It may be added here that calcium citrate is soluble to the extent of about 0.090 gm. per 100 c.c. of water at ordinary room temperatures, while the amount of calcium citrate formed by such a reaction in milk containing 3.2 gm. of casein per 100 c.c. has been found by us to be not over 0.066 gm. per 100 c.c.; it all, therefore, remains in solution passing through the filter into the serum.

If, then, we determine in milk the amount of soluble and insoluble calcium and then add to the milk sodium citrate, filtering and determining the calcium in the filtrate, there should, on the basis of our hypothesis, be an increased amount of calcium in the filtrate, showing how much calcium is transferred from the form of calcium caseinate to the form of calcium citrate.

EXPERIMENTAL WORK.

In carrying out our experimental work, we proceeded in the following manner: Fresh separator skim-milk was used, in which the amounts of casein, soluble and insoluble calcium, magnesium and phosphorus were determined. To prevent bacterial action, 3 c.c. of 40 per ct. formaldehyde solution was added to each liter of milk. The milk was then divided into nine equal parts and to each part was added, in varying amounts, crystallized sodium citrate (containing 27.7 per ct. of water of crystallization), as indicated in the table given below. The milk was then allowed to stand long enough for the reaction to reach equilibrium. Each portion was then filtered through a Chamberland filter and the amounts of calcium, magnesium and phosphorus were determined in the filtered serum. The results are given in Table I. Experiments were also made to test the effect of sodium citrate on the curdling action of rennin, the results of which are given in Table II.

Attention is called to the following points in connection with a study of the results contained in Table I.

1. *Changes in solubility of calcium.*— In the columns headed " calcium " we give the amounts of soluble and insoluble calcium in the

original untreated milk used in the experiments, and then, following, the amounts of soluble and insoluble calcium in the milk after treatment with amounts of sodium citrate varying from 0.130 to 1.040 gm. per 100 c.c. of milk (equivalent to 0.55 to 4.40 grains per ounce of milk). As previously stated, the soluble calcium is the portion appearing in the serum after filtering the milk under pressure through a Chamberland filter, while the insoluble calcium is that which fails to pass through the filter. An examination of the figures in the table shows that the amount of soluble calcium in 100 c.c. of the original milk is 0.045 gm. and this increases quite uniformly after each addition of increasing amounts of sodium citrate, the insoluble decreasing in essentially the same amounts. The only interpretation of these results that we can give is that some of the calcium of the caseinate or phosphate in the milk has been replaced by the sodium of the added citrate in the manner already discussed.

2. *Changes in solubility of phosphorus.*— The question suggests itself as to whether or not the increase of soluble calcium may come from action of sodium citrate on the insoluble calcium phosphate in the milk, forming sodium phosphate and calcium citrate. An examination of the figures in the columns under " Phosphorus " shows that there is no increase of soluble phosphorus until we have added more than 0.520 gm. of sodium citrate per liter of milk (equivalent to 2.20 grains per ounce), an amount sufficient to prevent curdling and even with larger additions the increase of soluble phosphorus is relatively small. The increase of soluble calcium comes, therefore, largely from the calcium that is combined with casein in the milk.

3. *Changes in solubility of magnesium.*— Owing to the small amount of magnesium in milk, the observed increase of solubility is slight but is in the direction shown by calcium, which would be expected.

In Table II we give the results obtained by treating 100 c.c. of milk with 2 c.c. of rennet solution (Shinn's liquid rennet) at 37° C. (98.8° F.). The rennet test was applied to untreated milk and also to samples of milk containing the varying amounts of crystallized sodium citrate given in the table.

Inspection of the results in this table makes it obvious that the presence of sodium citrate in milk, even in small amounts, delays very markedly the time of rennet curdling, while increase of citrate increases the time required for curdling, until we reach a point (0.400 gm. per 100 c.c. of milk or 1.7 grains per ounce), where no curdling takes place under the conditions of our experiments. It should be stated, in addition, that the character of the curdled milk varied in a characteristic way with the amount of sodium citrate added. Sample 1, untreated milk, gives a firm curd; the treated samples give curd of increasing softness with increase of sodium citrate.

The experimental results embodied in Tables I and II show that when sodium citrate is added to normal milk, (1) the amount of

TABLE I.—EFFECT OF SODIUM CITRATE IN INCREASING SOLUBLE CALCIUM, ETC., IN MILK.

No. of Experiment.	Grams of sodium citrate per 100 c.c. of milk.	Equal to grains of sodium citrate per ounce of milk.	Effect of adding 2 c.c. of rennet solution to 100 c.c. of milk at 37° C. (98.8° F.).	Grams per 100 c.c.					
				CALCIUM.		PHOSPHORUS.		MAGNESIUM.	
				Soluble.	Insoluble.	Soluble.	Insoluble.	Soluble.	Insoluble.
1	0.0	0.0	Curdled	.045	.101	.056	.035	.010	.004
2	0.130	0.55	Curdled	.052	.094	.055	.036	.010	.003
3	0.260	1.10	Curdled	.054	.093	.055	.036	.010	.003
4	0.390	1.65	Curdled	.056	.090	.055	.036	.010	.003
5	0.520	2.20	Not curdled	.059	.087	.056	.037	.011	.002
6	0.650	2.75	Not curdled	.064	.082	.061	.030	.011	.002
7	0.780	3.30	Not curdled	.069	.078	.061	.030	.011	.002
8	0.910	3.85	Not curdled	.072	.074	.062	.029	.012	.001
9	1.040	4.40	Not curdled	.075	.071	.064	.027	.012	.001

soluble calcium increases; (2) this increase is largely due to reaction of sodium citrate with the calcium caseinate in the milk, forming sodium caseinate or a double salt (calcium-sodium caseinate) and calcium citrate; and (3) use of increased amounts of sodium citrate lengthens the time required for the milk to curdle with rennet action or entirely prevents the curdling.

TABLE II.—EFFECT OF SODIUM CITRATE ON THE CURDLING OF MILK BY RENNIN.

No. of Experi-ment.	Grams of sodium citrate added to 100 c.c. of milk.	Equal to grains of sodium citrate in 1 ounce of milk.	Amount of rennet solution used per 100 c.c. of milk.	Minutes required for milk to curdle.
1.......	0.0	0.0	2	6
2.......	0.050	0.20	2	7½
3.......	0.100	0.40	2	8½
4.......	0.150	0.65	2	11
5.......	0.200	0.85	2	31
6.......	0.250	1.00	2	37
7.......	0.300	1.25	2	47
8.......	0.350	1.50	2	62
9.......	0.400	1.70	2	Not curdled
10.......	0.450	1.90	2	Not curdled
11.......	0.500	2.10	2	Not curdled

CONCLUSION.

Without going into full details, it will answer our purpose here to offer, on the basis of the facts developed, the following explanations as to why the presence of sodium citrate in milk delays or prevents curdling by rennin: Knowing the amount of soluble calcium formed by addition of sodium citrate to milk and knowing also the amount of casein present, we are able to ascertain that, at the point at which the rennet solution fails to curdle the milk, we have in place of the calcium caseinate of the normal milk a double salt, calcium-sodium caseinate. This double salt, on addition of rennin, forms a calcium-sodium paracaseinate containing one equivalent of sodium; and this compound, owing to the presence of the sodium, is not curdled by rennin. The increasing softness of the curd, accompanying the addition of increasing amounts of sodium citrate, is due to the presence of increasing amounts of calcium-sodium paracaseinate. A point is finally reached when all the calcium caseinate becomes calcium-sodium paracaseinate and the milk fails to curdle.

II. THE USE OF SODIUM CITRATE FOR THE DETERMINATION OF REVERTED PHOS-PHORIC ACID.[1]

ALFRED W. BOSWORTH.

In 1871, Fresenius, Neubauer and Luck [2] published a method for the determination of reverted phosphoric acid in phosphates which involves the use of a solution of neutral ammonium citrate, specific gravity 1.09. This method, with a change in the temperature of the solvent, has been in constant use since that time [3] with no attempt by any one to give an explanation of the chemical reaction involved. It has been quite generally believed that the neutral ammonium citrate solution possesses a selective power which enables it to separate dicalcic-phosphate from tricalcic-phosphate. This is not true, for it has been found in this laboratory that 100 c.c. of the official ammonium citrate solution [3] is capable of dissolving 1.3 grams of precipitated tricalcic-phosphate in one-half hour at a temperature of 65° C. This dissolving of the tricalcic-phosphate is accompanied by a precipitation of calcium citrate.

This separation of calcium citrate led to the belief that the solvent action of the citrate solution was the result of a double decomposition started by the free phosphoric acid always present in an aqueous solution which is in contact with a solid phase composed of a phosphate.[4] This double decomposition might be indicated by the following:

$$CaHPO_4 + 2C_6H_5O_6(NH_4)_3 \longrightarrow (NH_4)_2HPO_4 + [C_6H_5O_6(NH_4)_2]_2Ca$$
$$Ca_3P_2O_8 + 6C_6H_5O_6(NH_4)_3 \longrightarrow 2(NH_4)_3PO_4 + 3[C_6H_5O_6(NH_4)_2]_2Ca$$

If appreciable amounts of calcium are taken into solution, calcium citrate will separate out.

$$3[C_6H_5O_6(NH_4)_2]_2Ca \longrightarrow 4C_6H_5O_6(NH_4)_3 + (C_6H_5O_6)_2Ca_3$$

A great deal of work has been done upon methods of making neutral ammonium citrate solutions and several such methods have been published. The fact that neutral ammonium citrate is very unstable and easily loses ammonia has not been sufficiently considered in this connection, however. Why should extreme care be taken to secure an absolutely neutral solution, if this solution is to lose ammonia when heated a few degrees above the room temperature? Most chemists who have used the neutral ammonium citrate solution know that ammonia is constantly given off during the half hour allowed for the solvent action to take place. The final result then,

[1] Read before the Association of Official Agricultural Chemists, Washington, D. C., Nov. 17, 1913, and published in the *Jour. Indust. Eng. Chem.*, 6 : 227.
[2] *Ztschr. Analyt. Chem.*, 10 : 133.
[3] U. S. Dept. Agr., Bur. of Chem., Bull. 107 (revised).
[4] Cameron and Hurst, *Jour. Amer. Chem. Soc.*, 26, 905.

is not the action of neutral citrate but rather the action of an acid citrate. There seemed to be no theoretical reason why a solution of sodium citrate should not be just as effective a solvent and it possesses two distinct advantages. It is a more stable salt and as the base in it is not volatile the solution would remain neutral throughout the whole operation. All trouble in securing a neutral solution would be eliminated, for a solution of citric acid could be neutralized with sodium hydroxide, using phenolphthalein as an indicator, or the neutral crystals of sodium citrate could be dissolved in water, and the solution made up to the required volume.

In order to learn what the action of a solution of sodium citrate might be, one was made which was of the same molecular concentration as the Official[1] ammonium citrate solution, i. e., 314 grams crystallized sodium citrate, $(C_6H_5O_6Na_3)_2 \cdot 11\ H_2O$, per liter. This solution was used to determine the amounts of insoluble and reverted phosphoric acid in several fertilizers, Thomas slag, ground bone, ground rock phosphate, dicalcic-phosphate, $CaHPO_4$, and tricalcic-phosphate, $Ca_3P_2O_8$. The results, together with those obtained by the use of the Official citrate solution, are given in the table. In connection with these figures, it is noticeable that the differences between the figures obtained with the two solutions are, in most cases, of the same magnitude as the variations in the figures obtained by different chemists working upon the same sample.[2] It is also interesting to know that Samples 5, 10 and 11, which show the largest differences, all contain bone. The duplicate determinations, in all cases, showed closer agreement with sodium citrate solution than with the Official citrate solution.

The Official method directs that the flask in which the reaction takes place should be loosely stoppered, during the time it is being maintained at 65° C., in order to prevent evaporation. The use of stoppers often results in the loss of a determination through the breaking of a flask. It is suggested that the flask be closed with a one-hole rubber stopper carrying an empty calcium chloride tube, 300 mm. in length, which will serve as a condenser. The use of such a condenser will not interfere with the shaking and it furnishes a vent which prevents the breaking of the flask.

The last column of the table shows the amounts of ammonia given off during the half hour of treatment with ammonium citrate solution prescribed by the Official method. This ammonia was caught in standard acid by means of an air current which was passed through the Erlenmeyer flask in which the solvent action was taking place. These figures seem to bear some relation to the difference given in the preceding column. By noticing the large amounts of ammonia given off by the Thomas slag, rock phosphate and ground bone

---

[1] U. S. Dept. Agr., Bur. Chem. Bull. 107 (revised).

[2] *Jour. Indust. Eng. Chem.*, 3 : 118 and 5 : 957. The differences between the extremes in these two cases are 1.23 per ct. and 0.90 per ct. respectively.

when treated with ammonium citrate at 65° C. for one-half hour an indication as to the reason for the liberation of the ammonia may be found. The fertilizing materials, after being extracted with water, leave a residue which, in most cases, contains alkaline material, alkaline phosphates, carbonates of calcium and magnesium and oxides of other elements. These all tend to drive off ammonia from the citrate solution.

TABLE III.—COMPARISON OF THE USE OF AMMONIUM CITRATE AND SODIUM CITRATE FOR THE DETERMINATION OF REVERTED PHOSPHORIC ACID.

	Total P_2O_5.	Water-soluble P_2O_5.	By Ammonium Citrate.		By Sodium Citrate.		Difference.	Cc. of N/10 NH_3 liberated in ½ hour at 65° C.
			Insol. P_2O_5.	Reverted P_2O_5.	Insol. P_2O_5.	Reverted P_2O_5.		
1................	10.63	6.18	1.75	2.76	2.61	1.84	0.91	14.9
2................	8.73	3.76	1.42	3.55	1.89	3.08	0.47	12.7
3................	9.58	6.50	0.76	2.32	1.11	1.97	0.35	10.9
4................	12.33	11.90	0.02	0.41	0.00	0.43	0.02	6.5
5................	14.59	1.21	4.01	9.37	9.07	4.31	5.06	12.5
6................	10.92	3.76	0.58	6.58	1.11	6.05	0.53	13.5
7................	11.18	8.73	0.34	2.11	0.66	1.79	0.32	16.0
8................	9.61	4.24	1.62	3.75	2.80	2.57	1.18	14.2
9................	7.31	0.95	2.59	3.77	4.58	1.78	1.99	10.5
10................	8.79	0.00	5.22	3.57	7.78	1.01	2.56	29.0
11................	19.91	1.84	6.34	11.73	14.89	3.18	8.55	23.0
12................	13.07	8.42	0.22	4.43	0.77	3.88	0.55	13.5
13................	11.69	4.33	3.68	3.68	4.15	3.21	0.47	16.2
Bone.............	20.95	0.00	13.36	7.59	15.82	5.13	2.46	14.0
Slag.............	17.57	0.00	9.40	8.17	15.69	1.88	6.29	65.0
Rock phosphate....	29.72	0.19	27.57	1.96	28.20	1.33	0.63	20.5
$CaHPO_4$ 1 gram taken......		......	0.00	......	0.00	......	0.00	3.5
$Ca_3P_2O_8$ 1 gram taken......		......	0.00	......	0.00	......	0.00	14.0
Ammonium citrate heated to 65° C.............			......	......	......	......	......	2.6
Ammonium citrate heated to 75° C.............			......	......	......	......	......	36.0

It is realized that the small amount of evidence presented in this paper does not settle the question as to the desirability of substituting sodium citrate for ammonium citrate in the determination of reverted phosphoric acid. The subject is simply brought forward at this time in order that those chemists who are interested may give it some thought.

TECHNICAL BULLETIN No. 35.   JULY, 1914.

New York Agricultural Experiment Station.

GENEVA, N. Y.

BACTERIA OF FROZEN SOIL.

H. JOEL CONN.

PUBLISHED BY THE DEPARTMENT OF AGRICULTURE.

BACTERIA OF FROZEN SOIL.

H. JOEL CONN.

SUMMARY.

1. The number of bacteria in frozen soil is generally larger than in unfrozen soil. This fact was first noticed by the writer in 1910-11 when connected with the Cornell University Agricultural Experiment Station. Recently it has been observed at a different locality and in two other soils, one very different from the first. It is true not only of cropped soil, as shown in the previous work, but also of sod and fallow soil.

2. The increase in number of bacteria after freezing is not due to the increase in soil moisture which usually occurs in winter.

3. The same increase in germ content may take place in potted soil, where there is no possibility that the bacteria are carried up mechanically from lower depths during the process of freezing.

4. The facts noted under the headings 2 and 3 make it very probable that the phenomenon is due to an actual growth of bacteria after the soil is frozen. Its influence on fertility is still an unknown factor.

5. The results given in this bulletin were obtained in a different laboratory and under quite different conditions from those previously reported, thus partly eliminating errors which might have crept in because of peculiarities of technique.

INTRODUCTION.

Until recently considerable attention was given to variations in numbers of bacteria in soil at different seasons and under various conditions; but it is now realized that quantitative work alone is of small significance, and this line of investigation has been largely abandoned. As a result we know very little about the seasonal variation in either kinds or numbers of the bacteria in soil. A study of qualitative seasonal variations has never been undertaken. Samples for quantitative study have often been secured from greenhouse soil; or, if from the field, they have not been taken frequently enough or for a sufficiently long period to yield complete data. The flora of winter soil, in particular, has scarcely been studied.

It was assumed, for a long time, that bacterial activity was almost, if not completely, absent while soil was frozen. This assumption was supported by both theoretical and experimental evidence.

Theoretically, it had been reasoned that bacteria could not make use of congealed water in their physiological activities. Experimentally, freezing had been shown to prevent the growth and eventually to kill some types of bacteria (as for example *B. typhosus*). Probably these reasons have been largely responsible for the small attention given to the bacterial flora of winter soil.

HISTORICAL.

Remy[1] was among the first to furnish information as to the seasonal variation in numbers of bacteria in field soil. His attention, however, was directed mainly toward the physiological functions of the bacteria, and he did not try to perfect his methods of quantitative study. It is perhaps for this reason that he found no great variation in the numbers, and that his highest count was not over 4,000,000 per gram. None of Remy's samples were taken during the winter.

Hiltner and Störmer,[2] in the course of some experiments designed to show the effects of CS_2-treatment and of fallowing, took samples throughout more than one year. A few of their samples were taken during the winter, but none of them were from soil that had been long frozen. Their highest count was made from soil that had been frozen only a few days before the sample was taken.

Fabricius and von Feilitzen[3] carried out quantitative studies on five soils throughout one vegetative season. They found that the germ content showed a close relationship to soil temperature, so far as they tested the matter; but they examined no winter samples.

Kruger and Heinze,[4] in the course of an investigation of fallow soil, took several samples for bacteriological study during one year, omitting the winter.

Engberding[5] took numerous samples during two seasons of plant growth from fallow and cropped plats, manured and unmanured plats. During most of the year the samples were taken at short intervals; but between October and March only two samples were taken, both of which were from the same plat. These two samples gave moderately high counts, although not so high as some of the others.

---

[1] Remy, Th. Bodenbakteriologische Studien. *Centbl. Bakt.* Abt. II, 8:657–662, 699–705, 728–735, 761–769. 1902.

[2] Hiltner, L., and Störmer, K. Studien über die Bakterienflora des Ackerbodens, mit besonderer Berücksichtigung ihres Verhaltens nach einer Behandlung mit Schwefelkohlenstoff und nach Brache. *Kaiserliches Gesundheitsamt, Biol. Abt. Land- u. Forstw.* 3:445–545. 1903.

[3] Fabricius, O., and von Feilitzen, H. Ueber den Gehalt an Bakterien in jungfräulichem und kultiviertem Hochmoorboden auf dem Versuchsfelde des Schwedischen Moorkulturvereins bei Flahult. *Centbl. Bakt.* Abt. II, 14:161–168. 1905.

[4] Kruger, W., and Heinze, B. Untersuchungen über das Wesen der Brache. *Landw. Jahrb.* 36:383–423. 1907.

[5] Engberding, D. Vergleichende Untersuchungen über die Bakterienzahl im Ackerboden in ihrer Abhängigkeit von äusseren Einflussen. *Centbl. Bakt.* Abt. II, 23:569–642. 1909.

In 1910, while associated with the Cornell University Agricultural Experiment Station at Ithaca, N. Y., the writer first called attention to the fact that the number of bacteria in frozen soil is greater than in unfrozen soil.[6] The following winter more data were collected and all were published in a second article[7] appearing in 1911. For the sake of comparison with the results obtained at Geneva during 1912–14, the Ithaca work must be summarized here.

TABLE I.— BACTERIAL COUNTS OF FIELD SOIL.
Samples taken at Ithaca, N. Y.

| DATE. | BACTERIA PER GRAM DRY SOIL, IN — | | | |
| | PLAT 4B | | PLAT 1B | |
	Center.	North end.	Center.	North end.
April 17, 1909	7,000,000	....	18,000,000	....
May 25, 1909	3,300,000	....	10,000,000	....
July 16, 1909	5,000,000	....	4,500,000	....
Sept. 16, 1909	10,000,000	....	*22,000,000	....
Oct. 8, 1909	9,200,000	....	8,000,000	....
Nov. 23, 1909	6,800,000	....	5,750,000	....
Jan. 21, 1910	10,000,000	....	13,000,000	....
Feb. 7, 1910	23,500,000	....	....	....
Feb. 26, 1910	33,000,000	....	27,000,000	....
Mar. 25, 1910	5,700,000	....	6,700,000	....
April 15, 1910	12,000,000	7,000,000	2,500,000	2,000,000
May 28, 1910	7,000,000	....	12,000,000	....
June 15, 1910	14,000,000	....	16,500,000	....
July 2, 1910	13,000,000	....	9,000,000	....
Aug. 20, 1910	8,500,000	....	15,000,000	....
Sept. 14, 1910	9,500,000	7,000,000	13,000,000	7,000,000
Oct. 12, 1910	....	4,000,000	....	6,000,000
Nov. 12, 1910	....	9,500,000	....	5,900,000
Nov. 30, 1910	....	7,000,000	....	7,000,000
Dec. 19, 1910	....	23,000,000	....	17,500,000
Jan. 4, 1911	....	15,500,000	....	20,000,000
Jan. 21, 1911	....	15,000,000	....	10,000,000
Feb. 6, 1911	....	†14,000,000	....	27,000,000
Feb. 22, 1911	....	23,000,000	....	21,000,000
Mar. 3, 1911	....	22,000,000	....	28,000,000
Mar. 29, 1911	....	†16,000,000	....	7,500,000
April 12, 1911	....	14,000,000	....	20,000,000

*This count is probably too high, as little but surface soil was included in the sample.
†These counts are inexact, because of the extremely rapid liquefaction of the plates.

[6] Conn, H. J. Bacteria in Frozen Soil. *Centbl. Bakt.* Abt. II, 28:422–434. 1910.
[7] Conn, H. J. Bacteria of Frozen Soil. II. *Centbl. Bakt.* Abt. II, 32:70–97. 1911.

Two field plats, in Dunkirk clay loam,[8] about twenty feet apart, and both cropped to millet, were sampled nearly thirty times during the two years. The counts are listed in Table I, taken from the second of the two articles already mentioned. These results are also plotted in Graph I. In this graph the relations between the bacterial count, the moisture content of the samples, and the average weekly temperature are shown.

The following points were brought out by these analyses: (a) The highest counts were all made while the soil was frozen. Out of seventeen counts made from frozen soil or from soil recently thawed, all were over 10,000,000 and only four under 15,000,000 per gram; while of the forty other samples only fourteen were over 10,000,000 and but five over 15,000,000. The highest count in unfrozen soil (22,000,000) was exceeded by seven of the winter counts. (b) During the winter the numbers of bacteria increased while the soil was well frozen, but tended to decrease when it thawed. (c) In general, increases and decreases in the numbers of bacteria accompanied rises and falls, respectively, in the moisture content. In January and February, 1911, however, a series of fluctuations occurred which seemed to be closely associated with the freezing and thawing of the soil, but which were plainly independent of changes in moisture content.

This relationship between the number of bacteria and the soil-moisture suggested that the increase in germ content during the winter might be due to the greater moisture content rather than to the difference in temperature. There are two ways in which this may be possible: the added water, even though frozen, may furnish better conditions for bacterial growth; or, provided this increase in moisture is due to a capillary rise of water from lower depths during freezing, it may carry bacteria up with it, thus increasing their numbers in the surface soil without actual multiplication. To test out this point a pot of soil was kept frozen with a constant moisture content of 40 per ct. The experiment was not carried on under the most satisfactory conditions, because warm weather made it necessary to employ a freezing mixture of snow and salt, which resulted in a much lower temperature than that usually found in the field. A further difference from field conditions was caused by the aeration which necessarily results from the ordinary method of filling a pot with soil.

The results of this pot experiment are summarized in Table II. They are inconclusive. An increase in germ content is shown while the soil is frozen; but it is a much smaller increase than that observed under field conditions, and is not followed by a decrease after thawing.

[8] The soil nomenclature of the Bureau of Soils has been used throughout this work. The soils mentioned are described in the Soil Surveys of Ontario and of Tompkins Counties, New York, published by this Bureau. They are all glacial lake-bottom soils; see Fairchild, New York State Museum Bul. 127, pp. 66. 1909.

The experiment was discontinued on April 11, too soon to make it certain that the numbers of bacteria would not, eventually, have returned to their original level.

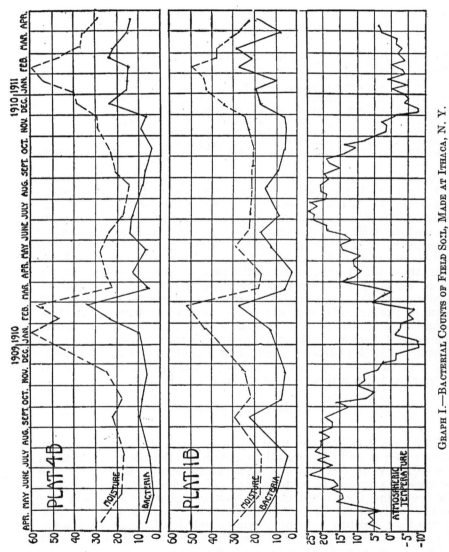

GRAPH I.—BACTERIAL COUNTS OF FIELD SOIL, MADE AT ITHACA, N. Y.

Numbers of bacteria expressed in millions per gram; moisture content in percentage, dry basis; temperature in degrees Centigrade, average per week.

The multiplication of bacteria in frozen soil, indicated by the analyses made at Ithaca, was so unexpected that some question has

TABLE II.— BACTERIAL COUNTS OF POTTED SOIL.
Samples taken at Ithaca, N. Y.

DATE.	Bacteria per gram dry soil.	Remarks.
1911.		
Jan. 4	5,000,000	Soil standing indoors, moisture 8 per ct.
Jan. 11	10,000,000	Soil indoors, moisture 40 per ct. since Jan. 4.
Jan. 16	12,000,000	Soil indoors, moisture 40 per ct. since Jan. 4.
Jan. 25	7,800,000	Soil frozen since Jan. 16; except on Jan. 21.
Feb. 11	10,700,000	Soil thawed Jan. 26–28; frozen since Jan. 29.
Mar. 20	16,000,000	Frozen artificially since thaw on Feb. 20.
Mar. 27	19,000,000	Still frozen, but not very stiff since Mar. 22.
April 4	21,000,000	Soil thawed since Mar. 27; kept indoors.
April 11	19,000,000	Still thawed; moisture 40 per ct.

been raised as to the correctness of the results. As yet, however, not much work has been done by others to test out the matter. Some unpublished work, carried on under the direction of W. M. Esten, of the Connecticut Agricultural College, has shown the germ content of soils to increase after freezing. Brown and Smith[9] recently made a study of bacteria in frozen soil, obtaining quantitative data from eight samples of soil. All of their counts are lower than those which were found in the present work, a fact which can be at least partially explained by their use of a different culture medium and of a shorter period (three days) of incubation. Although some of their counts from frozen soil were lower than others made before freezing, the highest count of all was from soil that had been the longest frozen. This fact is particularly interesting when we consider that the bacteria which show the most striking increase in numbers after freezing grow very slowly on the plates and are largely overlooked when a short period of incubation is used.

PRESENT WORK.

PLAN.

This work was planned to throw light upon the same two questions which it had been hoped to answer by the pot experiment in the earlier work (see page 6). The first question is whether the increase in numbers of bacteria may not be due merely to a rise of the organisms from lower depths, brought about by ascending currents of soil-water. The second is whether it is the low temperature or the high moisture content of winter soil that favors the bacteria. To answer these questions, soil was allowed to freeze in pots, so that its moisture content could be controlled and no water could rise

[9] Brown, P. E., and Smith, R. E. Bacterial Activities in Frozen Soils. Iowa Agr. Exp. Station, Research Bul. 4:158-184. 1912.

from below. Both aerated and unaerated soil were used in these pots, the unaerated soil having been obtained by digging a block from the field and transferring it directly to a pot. It was hoped by this means to see whether the failure of the previous pot experiment to show an appreciable increase in germ content could have been due to the unnatural aerated condition of the soil.

Parallel to these pot experiments, tests were made of the same soil (Dunkirk silty clay loam) in the field. In these tests, also, both aerated and unaerated soils were used. The unaerated soil was merely an undisturbed field plat that had been kept fallow since 1911, the upper two or three inches having been cultivated after every rain to preserve a dust mulch. There were two portions of aerated soil, one aerated in November, 1911, the other in November, 1913. The former portion, after aeration, was replaced within a large tile such as is used for sewer pipe, two feet in diameter and two feet long, buried in the field with only its flange above ground, and the soil so placed within it that the subsoil was below and the surface soil above in a layer of the same depth as occurs naturally in the field. This soil, like that of the unaerated plat, had been kept fallow since 1911. The second portion, aerated in 1913, consisted of surface soil alone and was replaced within a smaller cylinder, six inches deep and six inches in diameter, likewise sunk in the field.

A third series of tests was made in an entirely different soil, Dunkirk fine sand. Two spots were chosen, about fifty feet apart. One was in sod; the other at the edge of a strawberry field, in a spot that was practically free from any vegetation, either because it had been frequently cultivated or because of the poor quality of the soil. These tests were made to see whether the bacteria increase in numbers in a frozen sand as well as in a clay loam. The samples examined were unfortunately few in number.

METHODS.

The methods employed have been kept as nearly as possible the same as those used in the earlier work. Soil samples were regularly taken by boring to about six inches, although some of the winter samples were taken to a slightly less depth because of the difficulty in boring through frozen soil. The soil thus obtained was thoroughly mixed, by sifting, if dry enough, or by stirring, if muddy. A 0.5-gram portion of this sample was finally selected and shaken for two minutes with 100 cubic centimeters of sterile water in a stoppered flask. After this shaking, the suspension was further diluted, care being taken to keep the contents of the flask in motion when any of the suspension was withdrawn. One cubic centimeter of the proper dilution was finally added to each plate. The dilutions used were 1: 100,000 and 1: 200,000 or 1: 200,000 and 1: 500,000. Three or four plates of each dilution were always made.

Various media were tried in this work; but none ordinarily gave higher counts than the soil-extract gelatin described in the earlier articles. As a result, the figures chosen for publication are the counts obtained upon that medium and are comparable with those of the previous work. The composition of the medium is:

Gelatin . . . . .12 per ct.
Soil-extract . . . . .20 per ct.
Dextrose. . . . . .0.1 per ct.

Reaction adjusted to 0.5 per ct. normal acid to phenolphthalein.

In order to obtain the soil-extract, soil was heated for an hour at one atmosphere pressure, then mixed with an equal weight of cold water, allowed to stand over night, then boiled for half an hour and filtered.

The plates were incubated at a temperature of 17.5° to 18° C. for seven days before counting. In averaging the counts, consideration was taken of the fact that when over one hundred colonies appeared on a plate, overcrowding generally prevented the development of some of the bacteria. In such cases the greater dilution almost invariably gave the higher count, and the lower dilution was disregarded. Occasionally, however, the higher count was obtained from the lower dilution even though there were over one hundred colonies per plate, and then the counts of both sets of plates were averaged. If, on the other hand, there were less than one hundred colonies per plate, the difference between the counts obtained from the two dilutions was likely to be less than between those from parallel plates of the same dilution; so in this case the counts obtained from all the plates were averaged. In choosing the dilutions an attempt was made to obtain between fifty and one hundred colonies per plate in one or the other of the dilutions employed.

RESULTS OF POT WORK.

This work was planned, as already mentioned, for two purposes: to control soil moisture and to prevent the rise of bacteria from lower depths. In 1912 two pots, one aerated and the other unaerated, were prepared and tested occasionally during the following winter. The soil was frozen for such short periods, however, that the experiment was unsatisfactory, and it was repeated a second year, with four pots instead of two. The soil in two of these pots was aerated, in the other two unaerated. The two pots prepared the first winter were left uncovered, so that their moisture content rose and fell much the same as the field soil. During the second winter the four pots were kept covered, and their moisture content remained constant until the thaw in March, when melting snow managed to get beneath the covers.

The results of this work are given in Tables III and IV; while the results for 1913–14 are also plotted in Graphs II and III. The same graphs and tables show the moisture content of the samples (referred to dry basis). The atmospheric temperature (also shown in Table VII) is plotted in the two graphs. The atmospheric temperature was obtained by averaging the daily mean temperature for each week.

TABLE III.— BACTERIAL COUNTS OF POTTED SOIL.

Samples taken at Geneva, N. Y., 1912-13.

DATE.	AERATED SOIL.		UNAERATED SOIL.		Remarks.
	Moisture content	Bacteria per gram dry soil.	Moisture content.	Bacteria per gram dry soil.	
	Per ct.		*Per ct.*		
Nov. 16, 1912	19.5	37,000,000	19.5	25,000,000	Unfrozen.
Dec. 3, 1912	23	50,000,000	23.5	37,000,000	Unfrozen.
Feb. 11, 1913	18.5	44,000,000	19	27,500,000	Frozen 10 days.
Feb. 19, 1913	34	60,000,000	30	57,000,000	Frozen 18 days.
April 15, 1913	18.5	48,000,000	18	28,000,000	Thawed since Feb. 21.

The most significant samples taken in 1912–13 are those of February 19th, which were from well-frozen soil. The analyses show a decided increase in germ content over all counts from the soil while unfrozen. The only other samples of frozen soil taken were on February 11th, ten days after the freeze and apparently before the bacteria had begun to increase in numbers. The results, therefore, bear out previous work, but depend upon too few determinations to be conclusive. It is interesting to notice that increases and decreases in numbers of bacteria throughout the experiment accompany rises and falls in the moisture content.

The results of the work in 1913–14 are more conclusive. During this year eight samples were taken of soil that had been frozen at least two weeks. Of these, all but one gave strikingly higher counts than those obtained in the fall or spring. The results plainly cannot be ascribed to changes in moisture content, for no increase in the latter occurred until the final thaw in March.

Considering both years' work together, we find that nine out of twelve samples of frozen soil were abnormally high in germ content. Of the three that were no higher than normal, two were taken so soon after the freeze that the bacteria probably had not had time to increase in numbers. The chief significance of this experiment lies in the fact that the possibility of bacteria rising from lower depths during the process of freezing was excluded, thus showing

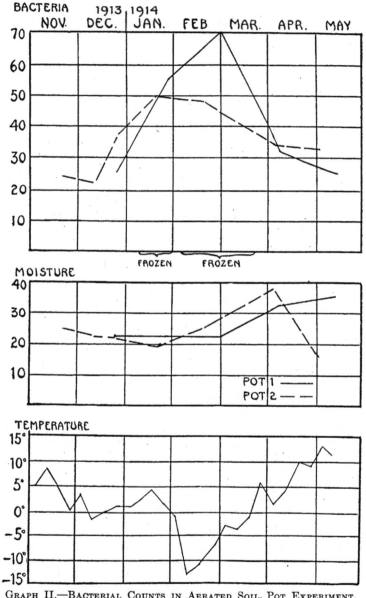

GRAPH II.—BACTERIAL COUNTS IN AERATED SOIL, POT EXPERIMENT, 1913-14.

Bacteria in millions per gram; moisture content in percentage, dry basis; temperature in degrees Centigrade, average per week.

TABLE IV.— BACTERIAL COUNTS OF POTTED SOIL.

Samples taken at Geneva, N. Y., 1913–14.

DATE.	AERATED SOIL.				UNAERATED SOIL.				Length of time frozen.
	POT 1.		POT 2.		POT 1.		POT 2.		
	Moisture content.	Bacteria per gram dry soil.	Moisture content.	Bacteria per gram dry soil.	Moisture content.	Bacteria per gram dry soil.	Moisture content.	Bacteria per gram dry soil.	
1913.	*Per ct.*		*Per ct.*		*Per ct.*		*Per ct.*		
Nov. 24	....	....	25	24,000,000	22	21,000,000	23	24,000,000	*
Dec. 11	....	....	22	22,000,000	21.5	16,000,000	....	....	3 days.
Dec. 26	22	26,000,000	22	37,000,000	....	....	....	....	*
1914.									
Jan. 24	....	....	19.5	50,000,000	17	24,000,000	....	....	15 days.
Jan. 28	22	56,000,000	....	....	....	....	24.5	60,000,000	19 days.
Feb. 20	....	....	24.5	49,000,000	26.5	58,000,000	....	....	15 days.
Feb. 28	22	71,000,000	....	....	....	....	21	42,000,000	23 days.
April 3	....	....	38.5	34,000,000	37	32,000,000	....	....	*
April 7	33.5	30,500,000	....	....	....	....	27	29,500,000	*
May 2	....	....	18.5	32,000,000	18	16,000,000	....	....	*
May 14	36	26,500,000	....	....	....	....	22	25,000,000	*

* Unfrozen.

conclusively that the increase in germ content could not have been due to this cause.

RESULTS OF FIELD WORK.

Dunkirk silty clay loam.— The analyses of this soil (the same type as used in the pot experiments) were made during three successive winters; but only the results secured in 1913–14 have sufficient meaning to justify publication in detail. During 1911–12 no constant temperature was available for use in the incubation of the plates, a condition which rendered the results unreliable. Seven samples of frozen soil were taken during that winter, of which two showed between thirty-five and forty million bacteria per gram, counts which are much higher than ordinarily obtained from this soil when in an unfrozen condition. In the winter of 1912–13 the soil was unfrozen except for two short periods, and only six samples were taken, of which only two were obtained as much as two weeks after a freeze. One of these two counts reached the striking figure of 55,000,000 bacteria per gram.

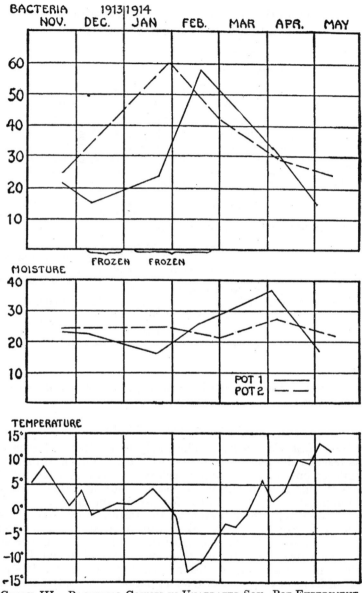

GRAPH III.—BACTERIAL COUNTS IN UNAERATED SOIL, POT EXPERIMENT, 1913-14.

Bacteria in millions per gram; moisture content in percentage, dry basis; temperature in degrees Centigrade, average per week.

TABLE V.— BACTERIAL COUNTS OF FIELD SOIL.

Samples of Dunkirk silty clay loam.

DATE.	UNAERATED.		AERATED, 1911.		AERATED, 1913.		Remarks.
	Moisture content.	Bacteria per gram dry soil.	Moisture content.	Bacteria per gram dry soil.	Moisture content.	Bacteria per gram dry soil.	
	Per ct.		Per ct.		Per ct.		
Nov. 24, 1913	....	....	....	....	23.5	40,000,000	Unfrozen.
Nov. 26, 1913	21.5	16,000,000	23.5	20,000,000	....	....	Unfrozen.
Dec. 15, 1913	19.5	11,000,000	22.5	13,500,000	25.5	22,000,000	Unfrozen.
Jan. 16, 1914	31	19,000,000	33.5	18,000,000	37.5	35,000,000	Frozen 9 days.
Jan. 30, 1914	24.5	33,000,000	23	34,000,000	....	....	Thawed 1 day.
Feb. 7, 1914	27.5	30,000,000	28	31,000,000	....	....	Partially frozen.
Feb. 26, 1914	31	38,000,000	....	....	30	59,000,000	Frozen 21 days.
April 15, 1914	20	21,000,000	20.5	23,000,000	23.5	32,000,000	Unfrozen.
April 29, 1914	20.5	16,000,000	20	13,000,000	20.5	22,000,000	Unfrozen.

The results for 1913–14 are worth giving in detail. During this year the weather was more favorable, samples were taken more frequently, and laboratory conditions were well controlled. As already mentioned, three series of tests were made: one from undisturbed field soil, fallow since 1911; one from aerated soil replaced in the field in November, 1911, and held fallow; and the third from soil aerated in November, 1913, and then replaced in the field. The results of the analyses are given in Table V. In Graph IV they are plotted, together with the moisture content of the samples and the average atmospheric temperature per week.

Seven of these counts were made from frozen soil, and two others from soil that had been thawed only twenty-four hours. All but two of these nine counts were above thirty million; and these two were taken only nine days after the first freeze of the winter. None of the counts made from unfrozen soil were above thirty million except two, both from the soil aerated in 1913, one on November 24th, immediately after aeration, and the other on April 15th, when this soil was found to be unusually moist and to have a musty odor because of a heavy piece of burlap that had been accidentally allowed to lie over it since the thaw. These exceptions occurred under such abnormal conditions that they are easily understood and do not affect the conclusion that the number of bacteria usually found in frozen soil is greater than in the same soil under ordinary summer conditions.

16

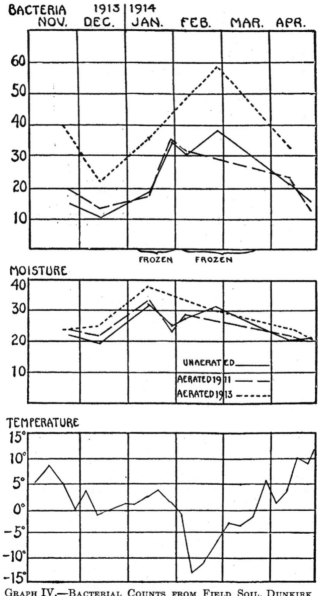

BACTERIA
NOV. | 1913 DEC. | 1914 JAN. | FEB. | MAR. | APR.

FROZEN   FROZEN

MOISTURE

UNAERATED _____
AERATED 1911 — — —
AERATED 1913 ------

TEMPERATURE

GRAPH IV.—BACTERIAL COUNTS FROM FIELD SOIL, DUNKIRK
SILTY CLAY LOAM.

Bacteria in millions per gram; moisture content in percentage,
dry basis; temperature in degrees Centigrade,
average per week.

Dunkirk fine sand.— The results of several tests in Dunkirk fine sand are given in Table VI. Three of the four samples of this sandy soil taken while frozen show distinct evidence of an increase in the germ content similar to that occurring in clay. The one sample from the cultivated portion taken on February 27th, shows the extremely high count of 95,000,000 per gram. This is a very interesting fact

TABLE VI.— BACTERIAL COUNTS OF FIELD SOIL.
Samples of Dunkirk fine sand.

DATE.	SOD.		CULTIVATED SOIL.		Remarks.
	Moisture content.	Bacteria per gram dry soil.	Moisture content.	Bacteria per gram dry soil.	
	Per ct.				
April 29, 1913	16	10,000,000	......	....	Unfrozen.
Nov. 18, 1913	11	8,000,000	9.5	9,000,000	Unfrozen.
Dec. 1, 1913	10	8,800,000	8	7,500,000	Unfrozen.
Feb. 27, 1914	34.5	57,000,000	29	95,000,000	Frozen 22 days.
Mar. 14, 1914	22	8,000,000	33	26,000,000	Frozen 37 days.
May 9, 1914	10	11,000,000	11	9,000,000	Thawed 2 months.

when the ordinary low count in this soil under normal conditions is noticed. It is impossible to tell, however, whether such a great increase in numbers is usual in this soil or is an isolated case to be found only in this one sample.

DISCUSSION.

In general, the results here obtained verify the previous results but do not explain them. They show that bacteria apparently increase in numbers in frozen soil when all possibility of their being carried up from lower depths is excluded. This increase seems to depend on the low temperature or upon freezing rather than upon an increase in soil moisture.

There is still some possibility, however, that the increase may not be an actual multiplication. Bacteria probably occur in the soil, to some extent, in masses that are too firmly bound together to be broken up by the shaking given the soil when it is diluted for plating. In such a case each colony developing on the plates represents either an isolated organism or an aggregate of two or more bacteria. It is entirely possible that the freezing may break up these masses and thus increase the plate count without adding to the actual number of bacteria present in the soil. It is very difficult to obtain any evidence bearing on this point. The only evidence so far obtained is the fact that bacteria have generally been found to reach their highest numbers two or three weeks after freezing, while if the increase were actually due to a breaking-up of the bacterial masses in the soil, the process ought to stop as soon as the soil is completely frozen. Moreover, if this be the true explanation, it is extremely unlikely that the count immediately after the thaw would be so nearly the same as it was before the freeze.

If this increase in germ content is due to an actual multiplication, the results seem to imply, on first thought, that soil organisms are able to use congealed water in their physiological activities. The soil-temperature, however, a few inches below the surface is seldom much below freezing in this climate, and the denser portions of the soil solution may never freeze. In these unfrozen portions, the growth of certain kinds of bacteria may be favored. This possibility was mentioned in the writer's first publication on the subject (1910).[10] Later it was further discussed by Brown and Smith.[11]

That certain bacteria can multiply at temperatures as low as this has been recognized for some time, cold storage conditions having been found insufficient to prevent all bacterial growth. It is strange, however, that low temperatures should seem to be more favorable than higher ones for the soil flora. The optimum temperature for nearly all of the soil bacteria which the writer has isolated has proved to be between 20° and 30° C. These bacteria show greater differences, however, in respect to their minimum temperature of growth. This fact is the basis of the theory already advanced to explain the increase in numbers while the soil is frozen. If we assume that there are two hostile classes of bacteria in the soil, one able to grow at temperatures below freezing, the other with its minimum temperature considerably higher, it is plain that sufficiently low temperatures would suppress one class and allow the other to increase with more than normal rapidity, even though these same organisms prefer higher temperatures in pure culture.

This theory ought not to be difficult to prove. If a distinct difference could be shown between the kinds of bacteria in frozen soil and those found when unfrozen, it would make this explanation seem extremely probable. No such difference, however, has been found. As the writer has elsewhere remarked,[12] there is a surprising similarity between the predominating bacteria found in different soils and in the same soil at different seasons. It is nevertheless possible, considering the crudity of present methods for classifying bacteria, that differences exist which have not been detected.

---

[10] See footnote 6.
[11] See footnote 11.
[12] Conn, H. J. The Distribution of Bacteria in Various Soil Types. *Journ. Amer. Soc. Agron.* 5:218–221. 1913.

TABLE VII.— ATMOSPHERIC TEMPERATURE, 1912–14.

1912.		Average temperature, degrees Centigrade.	1913		Average temperature, degrees Centigrade.
November	1– 8	8.5	August	1– 8	21.5
	9–15	9		9–15	23.5
	16–23	5.5		16–23	20
	24–30	0.5		24–31	19
December	1– 8	4	September	1– 8	24
	9–15	—1		9–15	24.5
	16–23	0		16–23	18
	24–31	1		24–30	16
1913.			October	1– 8	15
January	1– 8	1		9–15	13.5
	9–15	—2		16–23	11.5
	16–23	4		24–31	8.5
	24–31	1.5	November	1– 8	5
February	1– 7	—8		9–15	8.5
	8–14	—7		16–23	5
	15–21	1		24–30	0.5
	22–28	—4	December	1– 8	4
March	1– 8	—6		9–15	—1
	9–15	6.5		16–23	0
	16–23	5		24–31	1
	24–31	5.5	1914.		
April	1– 8	6	January	1– 8	1
	9–15	9		9–15	2.5
	16–23	10		16–23	4
	24–30	14.5		24–31	1.5
May	1– 8	18	February	1– 7	—1
	9–15	9.5		8–14	—13
	16–23	14.5		15–21	—10.5
	24–31	13.5		22–28	—7
June	1– 8	17	March	1– 8	—3
	9–15	18		9–15	—3.5
	16–23	19.5		16–23	—1
	24–30	23.5		24–31	6
July	1– 8	24	April	1– 8	1.5
	9–15	19.5		9–15	4
	16–23	21.5		16–23	10
	24–31	24		24–30	9.5
			May	1– 8	13.5
				9–15	12

Another explanation, offered by Russell,[13] is quite similar to this, but assumes that the hostile organisms suppressed by the low temperatures are not bacteria, but larger organisms, probably Protozoa. Russell, indeed, thinks it probable that the bacteria in soil are normally held in check by these protozoa; and that only after soil has been heated, frozen, dried, or treated with antiseptics, can the bacteria multiply to the greatest possible numbers. This theory is

---

[13] Russell, E. J. The Effect of Partial Sterilization of Soil on the Production of Plant Food. *Jour. Agr. Sci.* 5:152–221. 1913.

likewise unsupported by any direct evidence. It is a particularly hard theory either to prove or to disprove because of the difficulty in determining whether protozoa live in the soil in their active state.

If Russell's theory is correct, the increase in germ content which takes place in frozen soil is closely related to that which has been shown to occur in partially sterilized soil. Until recently the best supported explanation of the latter phenomenon was that the treatment necessary to effect partial sterilization disturbed the equilibrium of the soil bacteria and as a result allowed certain kinds to multiply abnormally. Now that Russell has proposed his protozoan theory, opinion is divided. Whichever explanation is the more probable, it is possible that the rapid increase in numbers of bacteria in partially sterilized soil and their multiplication in frozen soil may be due to similar causes. The improved crop-yields in the former case raise the question as to whether the increased germ content in the latter case has any practical importance. If the bacteria that multiply during the winter are favorable to plant growth, a cold winter may have a more beneficial effect on following crops than a warm one. This question leads into the unsolved problem of seasonal variation among soil bacteria. It shows the necessity of knowing what kinds of bacteria predominate at different seasons, and what influence each kind has upon plants.

TECHNICAL BULLETIN No. 36.   JULY, 1914.

New York Agricultural Experiment Station.

GENEVA, N. Y.

ORGANIC PHOSPHORIC ACIDS OF WHEAT BRAN.

I. CONCERNING THE ORGANIC PHOSPHORIC ACID COMPOUND OF WHEAT BRAN. II.

(Ninth Paper on Phytin.)

II. CONCERNING INOSITE MONOPHOSPHATE, A NEW ORGANIC PHOSPHORIC ACID OCCURRING IN WHEAT BRAN.

(Tenth Paper on Phytin.)

R. J. ANDERSON.

PUBLISHED BY THE DEPARTMENT OF AGRICULTURE.

ORGANIC PHOSPHORIC ACIDS OF WHEAT BRAN.

R. J. ANDERSON.

SUMMARY.

I. The results of a more extensive investigation of the organic phosphoric acid compound of wheat bran completely confirm the results previously published in Technical Bulletin 22.

Again it has been impossible to isolate any salts of phytic acid or inosite hexaphosphate.

The amorphous barium salts obtained agree in composition with those previously reported.

It appears probable, however, that these amorphous salts are not homogeneous but that they are mixtures of salts of various organic phosphoric acids. The isolation of definitely homogeneous compounds from this mixture has not succeeded.

Attention is called to the rather large content of oxalates in the crude organic phosphoric acid compound and also to the high percentage of inorganic phosphate contained in wheat bran.

II. A previously unknown organic phosphoric acid, inosite monophosphate, $C_6H_{13}O_9P$, has been isolated in beautiful crystalline form from wheat bran.

All the salts of this acid, with the exception of the lead salt, are very soluble in cold water. The alkaline earth salts are not precipitated with ammonia, differing in this respect from other known organic phosphoric acids as well as from ordinary phosphoric acid.

I. CONCERNING THE ORGANIC PHOSPHORIC ACID COMPOUND OF WHEAT BRAN. II. ·

(*Ninth Paper on Phytin.*)

INTRODUCTION.

It has been shown in a preliminary report[1] from this laboratory that the composition of the organic phosphoric acid isolated from wheat bran is different from that of phytic acid or inosite phosphoric acid which is present in other grains and seeds. Patten and Hart,[2] who first investigated this substance from wheat bran, came to the conclusion that it was identical with the " anhydro-oxymethylene diphosphoric acid " of Posternak. The analysis which they report

---

[1] *Journ. Biol. Chem.* 12: 447 (1912); and N. Y. Agric. Exp. Sta. Tech. Bull. 22. (1912).

[2] *Amer. Chem. Journ.* 31: 566. 1904.

of the acid preparation which they had isolated was in close agreement with the calculated composition of the above acid.

We suggested in our earlier report[3] that the acid analyzed by the above authors must have been contaminated with inorganic phosphoric acid because wheat bran contains appreciable quantities of inorganic phosphate and by their method of isolation both the organic and the inorganic phosphates would be precipitated at the same time.

The substances which we prepared from wheat bran and analyzed[4] were free from inorganic phosphate, i. e., they gave no precipitate with ammonium molybdate solution. In order to obtain preparations free from inorganic phosphate we found it necessary to precipitate the substance repeatedly from very dilute hydrochloric acid with about an equal volume of alcohol. By this process inorganic phosphates are removed because they are more soluble in the dilute acid-alcohol mixture than are the salts of the organic phosphoric acid.

In this way we obtained a crude preparation of the organic phosphorus compound which was readily soluble in cold water. Combined with it were various bases, calcium, magnesium, potassium, sodium, etc., and also some substance which contained nitrogen. By treating this crude substance in aqueous solution with barium hydroxide the above mentioned bases as well as the nitrogen-containing compound were eliminated. The resulting insoluble barium salts were amorphous and could not be obtained in crystalline form. These salts did not have the composition of barium phytates but agreed approximately with the formulas $C_{25}H_{55}O_{54}P_9Ba_5$ and $C_{20}H_{45}O_{49}P_9Ba_5$; from both salts an acid was obtained which approximately agreed with the formula $C_{20}H_{55}O_{49}P_9$. All the various preparations which were prepared in various ways agreed with the above formulas but since the substances were all amorphous we stated particularly[5] that, " The empirical formulas suggested in this paper are of course purely tentative."

Although we begged to reserve the further study of this organic phosphoric acid compound as well as the nitrogen-containing substance, Rather[6] in a recent paper reports some work on the same subject. This author had isolated some crude acid preparations from cottonseed meal and wheat bran from which silver salts were prepared. It is claimed that these silver precipitates are pure homogeneous compounds and that they are salts of an organic phosphoric acid of the formula $C_{12}H_{41}O_{42}P_9$. Since these results did not agree with those of any previous investigator in this field,

---

[3] Loc. cit.
[4] Loc. cit.
[5] Loc. cit.
[6] *Journ. Amer. Chem. Soc.* 35:890 (1913); and Texas Agric. Exp. Sta. Bull. 156. (1913).

the author concludes that his results are the only correct ones; owing to his superior method of isolation and purification, "purer" products had been obtained and he proposes the formula $C_{12}H_{41}O_{42}P_9$ as the correct one for phytic acid or inosite phosphoric acid.

We have already shown[7] that the above author is in error in respect to the composition of the acid in cottonseed meal. The carefully purified and many times recrystallized barium salts which we prepared from the acid from cottonseed meal had the composition of acid barium salts of inosite hexaphosphate, $C_6H_{12}O_{24}P_6Ba_3$ and $(C_6H_{11}O_{24}P_6)_2Ba_7$. The free acid prepared from these salts corresponds to inosite hexaphosphate, $C_6H_{18}O_{24}P_6$, and not to an acid of the formula $C_{12}H_{41}O_{42}P_9$.

The silver precipitates which we prepared from the above inosite hexaphosphate were pure white amorphous substances and very slightly sensitive to light. We showed,[8] however, that these silver precipitates are not homogeneous salts of inosite hexaphosphate but mixtures of more or less acid salts of the above acid.

In again taking up the investigation of the organic phosphoric acid compounds of wheat bran we have first of all critically repeated our former work. The results completely confirm those reported in our earlier paper.[9] We then repeated the work of Rather following his method of isolating the crude acid as closely as possible. The acid preparation obtained in this way was divided into two parts: One portion was used for the preparation of the silver salt as described by the above author; the other portion was transformed into the barium salt and purified in accordance with our previous method.

The barium salts which were obtained in this way were found to agree very closely in composition with those previously reported, viz.: $C_{20}H_{45}O_{49}P_9Ba_5$ and not with salts of the acid $C_{12}H_{41}O_{42}P_9$.

The silver precipitates which were obtained from the crude acid varied in composition according to the method of preparation but in one case the substance had approximately the composition stated by Rather. A simple examination of these silver precipitates quickly revealed the fact, however, that they were not "pure homogeneous salts" of an organic phosphoric acid of the formula $C_{12}H_{41}O_{42}P_9$ as claimed by the above author, but that they were largely contaminated with inorganic silver phosphate — varying from 42 to 90 per ct.

In our first report[10] on this subject we called attention to the fact that wheat bran extracts contain relatively much inorganic phos-

---

[7] *Jou n. Biol. Chem.* 17: 141 (1914); and N. Y. Agric. Exp. Sta. Tech. Bull. 32. (1914).r

[8] Ibid., p. 149.

[9] *Journ: Biol. Chem.* 12: 447 (1912); and N. Y. Agric. Exp. Sta. Tech. Bull. 22. (1912).

[10] Loc. cit.

phate. This part of our paper seems to have escaped the attention of Rather. This author, like Patten and Hart, has made no provision for eliminating inorganic phosphate in his method of isolating the organic phosphoric acid.

Since inorganic and organic phosphoric acids are both present in the crude acid prepared in accordance with the methods of the above authors naturally the silver precipitates obtained from such an acid neutralized with ammonia must contain both inorganic and organic silver phosphates because both are only slightly soluble in neutral aqueous solution.

Although the silver precipitates obtained from cottonseed meal and wheat bran may have approximately the same composition it is surprising that anyone could consider them identical; for so far as the most obvious physical property, viz., appearance, is concerned, they are entirely dissimilar. The silver precipitates from the inosite hexaphosphate from cottonseed meal are, as already mentioned, of pure white color and they are very slightly affected by light. The silver precipitates obtained from the acid from wheat bran, on the other hand, are only white at the moment of precipitation. These substances are either extremely sensitive to light or else the silver becomes reduced for the color rapidly darkens and finally turns quite black. Even when working under careful exclusion of direct light we have been unable to obtain a white silver preparation from the wheat bran compound.

While the amorphous barium salts prepared, as will be described later, from the organic phosphorus compound of wheat bran, show a close agreement in composition with those reported previously[11] we do not believe that they are homogeneous compounds. We have been able to separate these amorphous precipitates into several fractions some of which were semi-crystalline, but the composition was not constant. In no case, however, have we been able to obtain a trace of a salt having the composition of inosite hexaphosphate.

From the results which we have obtained it appears probable that these amorphous barium precipitates are mixtures, probably of various organic phosphoric acids. Some of these are undoubtedly lower phosphoric acid esters of inosite but it is possible that phosphoric acid esters of other carbohydrates are also present.

Neither Patten and Hart nor Rather mention the presence of oxalic acid in the preparations from wheat bran which they examined. It would seem from our results, however, that the crude substance obtained by precipitating a dilute hydrochloric acid extract of wheat bran with alcohol contains rather large quantities of oxalates.

The removal of this oxalate presented greater difficulties than the elimination of inorganic phosphate. As a barium salt oxalic acid is precipitated at every stage along with the salts of the organic phos-

[11] Loc. cit.

phoric acid. It is likewise carried down in the precipitate obtained with copper acetate because copper oxalate is very slightly soluble. The complete removal of the barium oxalate from the other mixture of barium salts was finally secured by allowing it to crystallize out from a very dilute hydrochloric acid solution of the mixed salts.

If the name " phytin " is to be applied to certain salts of inosite hexaphosphate then it is evident that wheat bran does not contain phytin for we have been unable to isolate any salt of this acid from this material.

It appears more probable that wheat bran contains several different organic phosphoric acids. So far we have been able to identify only one of these acids, viz., inosite monophosphate, a substance which will be described in a succeeding paper.

EXPERIMENTAL PART.

ISOLATION OF THE CRUDE ORGANIC PHOSPHORUS COMPOUND FROM WHEAT BRAN.

A larger quantity of the crude natural organic phosphorus compound was prepared by precipitating a 0.2 per ct. hydrochloric acid extract of wheat bran with alcohol exactly as before. From 25 kilograms of wheat bran we obtained 222 grams of the crude compound as a nearly white amorphous powder. This substance had the following composition: $C = 20.21$; $H = 3.54$; $P = 13.45$; $Mg = 8.20$; $K = 5.23$; $Na = 2.56$; $Ca = $ trace. Nitrogen was present but it was not determined. The substance was practically free from inorganic phosphate as it gave only a trace of a yellow precipitate on warming its nitric acid solution to 65° for a longer time with ammonium molybdate.

Of the above substance, 50 grams were suspended in a small amount of water and dissolved by the addition of a little hydrochloric acid. After diluting with water it was precipitated by adding barium hydroxide in excess. It was then filtered and thoroughly washed in water.

PREPARATION OF THE NITROGEN-CONTAINING SUBSTANCE.

The filtrate from the above was freed from excess of barium hydroxide with carbon dioxide, filtered and evaporated in vacuum at a temperature of 40°–45° to small bulk and again filtered from a small amount of barium carbonate. The concentrated solution was then poured into about 500 c.c. of alcohol. It separated as a somewhat sticky mass which soon hardened. It was filtered, washed in alcohol and ether and dried in vacuum over sulphuric acid. It weighed 7 grams.

It was dissolved in a small amount of water and reprecipitated with alcohol, filtered, washed and dried as before. It was obtained as a nearly white amorphous powder. The substance was very soluble in water and it was free from chlorides and inorganic phos_phate.

It was analyzed after drying at 105° in vacuum over phosphorus pentoxide.

Found: $C = 39.22$; $H = 5.43$; $N = 14.26$; amino N by the Van Slyke method $= 0.999$; $Ba = 10.43$; organic phosphorus after decom_posing by the Neumann method $= 0.875$; it also contained small quantities of magnesium, potash and soda.

PREPARATION OF THE CRUDE BARIUM SALT OF THE ORGANIC PHOS_PHORUS COMPOUND.

The insoluble precipitate obtained with barium hydroxide from the 50 grams of the crude substance was again precipitated three times with barium hydroxide from 0.5 per ct. hydrochloric acid and then four times with alcohol from the same strength hydro-chloric acid. The product was then a white amorphous powder and it weighed after drying in vacuum over sulphuric acid 22 grams.

A preliminary experiment showed that when this substance was dissolved in a little 0.5 per ct. hydrochloric acid and then mixed with a concentrated solution of barium chloride and allowed to stand some crystalline precipitate separated. The whole of the above barium salt was therefore dissolved in the least possible amount of 0.5 per ct. hydrochloric acid; 5 grams of barium chloride dissolved in a little water was added and the whole allowed to stand for 24 hours. A heavy white crystalline powder had then separated. This was filtered and washed several times in water and finally in alcohol and ether and allowed to dry in the air. This crystalline substance was found later to consist principally of barium oxalate. It weighed 2.75 grams which is equal to 12.5 per ct. of the barium salt used.

The filtrate, after removing the above crystals, was precipitated with alcohol, filtered, washed in dilute alcohol, alcohol and ether and dried in vacuum over sulphuric acid. It was again dissolved in the minimum quantity of 0.5 per ct. hydrochloric acid, mixed with a solution of barium chloride and allowed to stand for 24 hours. A small amount of a crystalline precipitate had separated which was filtered off. The filtrate was again precipitated with alcohol, filtered, washed and dried as before. The dry substance was again dissolved in 0.5 per ct. hydrochloric acid, barium chloride added and allowed to stand for 24 hours. The solution remained perfectly clear and no precipitate had separated. The substance was then precipitated with alcohol, filtered and washed, again dissolved in 0.5 per ct. hydrochloric acid and reprecipitated with alcohol. It was finally filtered, washed in dilute alcohol, alcohol and ether and dried in vacuum over sulphuric acid. The product was a snow-white amorphous powder, free from chlorides and inorganic phosphate and the following results were obtained on analysis: C = 14.98; H = 2.46; P = 11.89; Ba = 31.64 per ct.

The substance was quite soluble in cold water. It was therefore rubbed up in a mortar with a small quantity of water in which the greater portion dissolved. After standing over night the insoluble portion was filtered off, washed in water, alcohol and ether and dried in vacuum over sulphuric acid. The water-soluble portion was precipitated with alcohol, filtered, washed in dilute alcohol, alcohol and ether and dried in vacuum over sulphuric acid.

After drying at 105° in vacuum over phosphorus pentoxide the following results were obtained on analysis:

The water-insoluble substance gave: C = 12.58; H = 2.02; P = 10.06; Ba = 40.62 per ct.

The water-soluble substance gave: C = 15.54; H = 2.95; P = 12.30; Ba = 30.24 per ct.

EXAMINATION OF THE CRYSTALLINE BARIUM OXALATE OBTAINED FROM THE DILUTE HYDROCHLORIC ACID SOLUTION OF THE ABOVE BARIUM SALT.

The substance was analyzed after drying at 105° in vacuum over phosphorus pentoxide and the following result obtained: C = 6.25; H = 0.81; P = 0.95; Ba = 55.32 per ct.

The phosphorus was present in organic combination. The ash was found to consist principally of barium carbonate. When some of the substance was heated with concentrated sulphuric acid a gas was liberated which caused a white precipitate of barium carbonate when led into barium hydroxide solution.

Judging by the analysis and reactions the crystalline substance was an impure barium oxalate mixed with some of the barium salt of the organic phosphoric acid.

The crude barium salt of the organic phosphoric acid first obtained contained therefore about 12 per ct. of barium oxalate.

That the substance was barium oxalate was further confirmed by the following experiments: It was recrystallized several times from hot dilute hydrochloric acid by partially neutralizing with ammonia. It was then transformed into silver salt as follows: the barium oxalate was dissolved in a little hot dilute nitric acid, diluted with water and silver nitrate added which caused a heavy white granular precipitate. This was filtered off, washed in water, alcohol and ether and dried in the air. This substance showed all the properties of silver oxalate, viz: it was very insoluble in dilute nitric acid and on heating the dry substance it exploded. It was, however, not free from phosphorus. It contained 68.41 per ct. of silver while silver oxalate contains 71.05 per ct. of silver.

The balance of the barium salt was then transformed into calcium oxalate and the latter was recrystallized many times from boiling dilute hydrochloric acid by nearly neutralizing with ammonia and acidifying with acetic acid. After purifying in this way 0.0826 gram of the substance was burned to constant weight in a platinum crucible. The calcium oxide remaining weighed 0.0310 grams. Calculated for the above quantity of $CaC_2O_4 + H_2O = 0.0316$ gram CaO.

The calcium oxide was dissolved in dilute nitric acid and tested with ammonium molybdate. A faint precipitate separated showing that some phosphorus remained. There appears, however, to be no doubt that the substance was a nearly pure calcium oxalate.

PREPARATION OF THE SILVER SALT OF THE ORGANIC PHOSPHORIC ACID.

The water-soluble barium salt analyzed on p. 9 was transformed into the silver salt as follows: 5 grams were suspended in water and decomposed with a slight excess of dilute sulphuric acid; the barium sulphate was filtered off and the filtrate neutralized to litmus with ammonia. Silver nitrate was added producing a pure white precipitate which however rapidly darkened in color. It was filtered, washed in water and dried in vacuum over sulphuric acid under exclusion of light. The substance was then a heavy dark-gray amorphous powder. It was free from all but traces of ammonia. For analysis it was dried at 105° in vacuum over phosphorus pentoxide under exclusion of light but it turned very dark in color.

Found: C = 9.49; H = 1.48; P = 7.89; Ag = 54.85 per ct.

The substance was free from inorganic phosphate.

As will be noticed it corresponds very closely in composition to the barium salt from which it was prepared and not to the compounds analyzed by Rather.

PREPARATION OF THE CRUDE ACID FROM WHEAT BRAN BY THE METHOD OF RATHER.

In the preparation of the acid, the directions of the above author were followed as closely as possible. The various operations may be briefly stated as follows: Wheat bran was digested in 0.2 per ct. hydrochloric acid for three hours with frequent stirring. It was then strained through cheesecloth, the residue was washed with water and again strained. The extract was centrifugalized and finally filtered. Copper acetate solution was added in excess and allowed to settle over night. The copper precipitate was freed from the mother-liquor as far as possible by the centrifuge and finally brought on the Buchner funnel and then washed several times in water. It was then suspended in water and decomposed with hydrogen sulphide, filtered and the filtrate evaporated to a thin syrup on the water-bath. This was dissolved in a small quantity of water and rendered strongly alkaline with ammonia and allowed to stand for 24 hours. The precipitate was filtered off and the filtrate evaporated on the water-bath until the excess of ammonia was driven off. It was then diluted with water and precipitated with barium chloride in excess, filtered and washed and suspended in cold water and decomposed with slight excess of dilute sulphuric acid. After filtering, the filtrate was neutralized with ammonia and again precipitated with barium chloride. These operations were repeated three times and after finally removing the barium the filtrate was precipitated with copper acetate — this was filtered and washed until the filtrate gave no reaction with barium chloride. The copper precipitate was then suspended in water and decomposed with hydrogen sulphide; filtered and the filtrate evaporated on the water-bath to a syrupy consistency. This syrup was poured into 1,600 c.c. of alcohol and allowed to stand for the precipitate to settle. It was then filtered and again evaporated on the water-bath until the alcohol was removed. The residue was the crude acid which was obtained as a brown-colored syrup. It was diluted with water to 100 c.c. in which it formed a slightly opalescent solution of faint aromatic odor. It was divided into two parts; 75 c.c. was used for the preparation of the barium salt; of the balance, 10 c.c. was used for the preparation of the silver salt.

PREPARATION OF THE SILVER SALT FROM THE CRUDE ACID.

The 10 c.c. of the above acid solution was diluted to 100 c.c. with water and ammonia added to alkaline reaction. The excess of ammonia was boiled off, the solution cooled and silver nitrate added which caused a voluminous yellow-colored amorphous precipitate. This was filtered, washed in water and dried in vacuum over sulphuric acid. The substance was very sensitive to light, and although the desiccator was kept in a dark place the dry salt was very dark in color.

The filtrate from above was quite acid in reaction. It was neutralized with ammonia when a further quantity of a yellowish precipitate came down. More silver nitrate was added and the precipitate filtered, washed and dried as before. This was also a very dark amorphous powder. The precipitates were free from all but traces of ammonia.

These silver precipitates were analyzed after drying at 105° in vacuum over phosphorus pentoxide. The first precipitate gave: C = 2.18; H = 0.53; Ag = 71.82; total phosphorus = 7.82; inorganic phosphorus = 5.61 per ct.

The inorganic phosphorus was determined as follows: the substance was suspended in cold water and dissolved by the addition of cold dilute nitric acid and the silver precipitated with hydrochloric acid. The filtrate was neutralized with ammonia, acidified with nitric acid, ammonium nitrate and ammonium molybdate added, and the whole kept at a temperature of 65° for one-half hour. The precipitate was then filtered and the phosphorus determined as magnesium pyrophosphate as usual.

The composition of the above precipitate is quite different from that reported by Rather. It will be noticed, however, that the inorganic phosphorus is equivalent to 75.64 per ct. of inorganic silver phosphate. The second precipitate gave the following: C = 0.98; H = 0.37; Ag = 74.37; total phosphorus = 7.15; inorganic phosphorus = 6.73 per ct. This substance accordingly contained 90.74 per ct. of inorganic silver phosphate.

PREPARATION OF THE BARIUM SALT FROM THE CRUDE ACID.

The 75 c.c. of the crude acid solution was diluted to 500 c.c. with water and precipitated with barium hydroxide in excess. The resulting barium precipitate was reprecipitated as before alternately with barium hydroxide four times and alcohol five times from 0.5 per ct. hydrochloric acid. After finally filtering, washing in dilute alcohol, alcohol and ether, and drying in vacuum over sulphuric acid, 7 grams of a snow-white amorphous powder was obtained. It was free from chlorides and inorganic phosphate.

It was analyzed after drying at 105° in vacuum over phosphorus pentoxide.

Found: C = 12.31; H = 2.21; P = 13.99; Ba = 33.45 per ct.

It will be noticed that after removing inorganic phosphate the composition of the barium salt does not agree with a compound of the formula $C_{12}H_{41}O_{42}P_9$ as proposed by Rather, but that the composition agrees closely with the compound $C_{20}H_{45}O_{49}P_9Ba_5$ which was described in the earlier paper (*loc. cit.*); calculated for the above: C = 11.79; H = 2.21; P = 13.71; Ba = 33.76 per ct.

SECOND PREPARATION OF THE CRUDE ACID FROM WHEAT BRAN BY THE METHOD OF RATHER.

A second lot of the acid was prepared from wheat bran by the same method as before except that the various concentrations were done in *vacuum* at a temperature of 40°–45° and not on the water-bath as the first time.

The silver and barium salts were prepared from the crude acid exactly as before.

The silver precipitate gave the following results on analysis after drying at 105° in vacuum over phosphorus pentoxide: C = 4.56; H = 0.77; Ag = 65.88; total phosphorus = 8.03; inorganic phosphorus = 3.15 per ct.

This approaches in composition the precipitates analyzed by Rather. However, as shown by the above content of inorganic phosphorus it contained 42.54 per ct. of inorganic silver phosphate.

The barium precipitate was free from chlorides and inorganic phosphate. It was analyzed after drying at 105° in vacuum over phosphorus pentoxide. Found: C = 11.66; H = 2.11; P = 14.14; Ba = 34.36 per ct.

This substance is therefore identical in composition with the first preparation reported above and agrees very closely with the compounds previously described.

It is clearly evident from the results recorded above that an acid, $C_{12}H_{41}O_{42}P_9$, such as described by Rather, does not exist, at least not in wheat bran or cottonseed meal. The alleged, "pure, homogeneous silver salts," analyzed by the above author must have been largely contaminated with silver phosphate and this simple impurity escaped his observation.

FURTHER PREPARATIONS OF THE BARIUM SALTS OF THE 'ORGANIC PHOSPHORUS COMPOUND OF WHEAT BRAN.

The 0.2 per ct. hydrochloric acid extract of wheat bran was precipitated by adding a concentrated solution of barium chloride. After settling, the precipitate was filtered and washed in dilute alcohol, alcohol and ether and dried in vacuum over sulphuric acid.

To the mother-liquor from above, barium acetate was added which caused a further precipitate. This was filtered, washed and dried in vacuum over sulphuric acid.

Total and inorganic phosphorus were determined in these precipitates as follows: Total phosphorus was determined after decomposing by the Neumann method. As inorganic phosphorus we consider the amount of phosphorus directly precipitated by ammonium molybdate from the nitric acid solution of the substance at a temperature of 65° for one-half hour.

Found in the precipitate produced with barium chloride:
Total phosphorus = 8.45 per ct.
Inorganic phosphorus = 1.55 per ct.

The result shows that 18.32 per ct. of the total phosphorus was present as inorganic.

Found in the precipitate produced with barium acetate:
Total phosphorus = 10.28 per ct.
Inorganic phosphorus = 8.63 per ct.

In this case 83.96 per ct. of the total phosphorus was present as inorganic.

The above barium precipitates were purified separately in the same way as before by repeatedly precipitating with barium hydroxide and alcohol alternately from 0.5 per ct. hydrochloric acid until pure white amorphous powders were obtained which were free from inorganic phosphate and which contained no bases except barium.

After drying at 105° in vacuum over phosphorus pentoxide the following results were obtained on analysis:

The preparation isolated from the barium chloride precipitate:
Found: C = 11.53; H = 2.10; P = 14.29; Ba = 34.60 per ct.

This substance has the same composition as the precipitates obtained from the previously isolated crude acid. These various preparations were therefore mixed and treated as will be described later.

The preparation isolated from the barium acetate precipitate:
Found: C = 15.22; H = 2.62; P = 12.22; Ba = 29.85 per ct.

Several other preparations were made from wheat bran in different ways. The composition varied considerably, however, as is evident from the figures given below.

One preparation gave the following:
C = 15.33; H=2.70; P = 11.61; Ba = 32.31 per ct.

Another gave:
C = 14.62; H=2.65; P = 12.84; Ba = 30.59 per ct.

A third preparation gave:
C = 13.94; H = 2.47; P = 13.10; Ba = 31.41 per ct.

A fourth gave:
C = 15.01; H = 2.67; P = 12.25; Ba = 30.94 per ct.

TREATMENT OF THE AMORPHOUS BARIUM SALT WITH COLD WATER.

The barium precipitates obtained from the crude acid as well as the substance isolated from the barium chloride precipitate which, as shown above, had the same composition, were mixed together. Total weight 22.5 grams. It was rubbed up in a mortar with 200 c.c. of cold water when the greater portion dissolved. The insoluble substance was filtered, washed in water, alcohol and ether and dried in vacuum over sulphuric acid. It was then treated with a second portion of 200 c.c. of cold water, again filtered, washed and dried. This insoluble residue weighed 5.75 grams. It had the following composition after drying at 105° as before:
Found: C = 9.52; H=1.53; P = 12.89; Ba = 41.79 per ct.

The filtrate from the above, containing the water-soluble portion of the barium salt, was heated to boiling. The solution turned first cloudy but gradually a white flocculent precipitate separated which soon assumed a granular form and settled to the bottom of the flask. This was filtered, washed in boiling water, alcohol and ether and dried in the air. Under the microscope the substance showed no definite crystalline structure but consisted of fine transparent globules. It weighed 1.6 grams. It was free from inorganic phosphate.

It was analyzed after drying at 105° in vacuum over phosphorus pentoxide.
Found: C = 9.15; H = 1.79; P = 13.57; Ba = 40.43; H_2O = 9.53 per ct.
    C = 8.96; H = 1.62.

The filtrate from the above was just neutralized with barium hydroxide. The white amorphous precipitate was filtered, washed in water, alcohol and ether and dried in vacuum over sulphuric acid. It weighed 17 grams. It did not contain inorganic phosphate. It was analyzed after drying at 105° as above:

Found: $C = 9.19$; $H = 1.42$; $P = 11.21$; $Ba = 47.86$ per ct.

CRYSTALLIZATION OF THE WATER-INSOLUBLE BARIUM SALT.

The water-insoluble barium salt previously analyzed, 5.75 grams, was dissolved in the minimum quantity of 0.5 per ct. hydrochloric acid and the solution heated to boiling. The solution remained perfectly clear. Alcohol was then added until the solution turned slightly cloudy. On scratching, the substance began to separate in a crystalline or granular form on the sides of the flask.

After standing over night the substance was filtered, washed in water, alcohol and ether and dried in the air.

The filtrate was precipitated with alcohol, filtered, washed and dried in vacuum over sulphuric acid. It was again dissolved in a small quantity of 0.5 per ct. hydrochloric acid, heated, alcohol added and allowed to stand as before when a further quantity of the same shaped crystalline or granular product was obtained. This was filtered, washed in water, alcohol and ether and dried in the air.

This product was a heavy snow-white crystalline or granular powder. Under the microscope it looked homogeneous and consisted of small transparent spherical globules. It was free from chlorides and inorganic phosphate. For analysis it was dried at 105° in vacuum over phosphorus pentoxide.

Found: $C=9.91$; $H=1.83$; $P=14.90$; $Ba = 34.75$; $H_2O = 12.48$ per ct.

This substance was recrystallized in the same manner when it separated in the same form as before. It was again analyzed after drying as above.

Found: $C = 10.06$; $H = 2.02$; $P = 15.30$; $Ba=33.36$; $H_2O=11.73$ per ct.

Since this substance separated in characteristic crystalline manner and did not show any appreciable change in composition on recrystallization one might believe that it was homogeneous. We have, however, been unable to obtain any other preparation having the same composition. Numerous preparations were made which separated in exactly the same manner and form and so far as appearance is concerned they looked identical, but on analysis widely varying results were obtained.

One product gave the following:

$C = 10.67$; $H = 2.00$; $P = 14.46$; $Ba = 35.02$; $H_2O = 11.97$ per ct.

Another was prepared and recrystallized three times when it gave the following:

$C = 11.60$; $H = 1.88$; $P = 12.48$; $Ba = 34.37$; $H_2O = 11.60$ per ct.

Another preparation gave the following:

$C = 11.26$; $H = 1.58$; $P = 10.67$; $Ba = 37.81$; $H_2O = 12.00$ per ct.

Judging by these results it is evident that wheat bran contains more than one organic phosphoric acid. It appears probable that several are present and that the solubility of the salts of such acids differs so slightly that their separation is very difficult. Until definitely homogeneous products can be separated from this mixture it seems futile to develop empirical formulas; for such may be calculated for every substance analyzed. The investigation is being continued.

The author wishes to express his appreciation and thanks to Dr. P. A. Levene of the Rockefeller Institute for Medical Research, New York, N. Y., and to Dr. Thomas B. Osborne of the Connecticut Agricultural Experiment Station, New Haven, Conn., for many valuable suggestions.

II. CONCERNING INOSITE MONOPHOSPHATE, A NEW ORGANIC PHOSPHORIC ACID OCCURRING IN WHEAT BRAN.*

(Tenth Paper on Phytin.)

INTRODUCTION.

In previous reports[1] we have shown that the crude organic phosphorus compound of wheat bran[2] can be separated into two portions by treating it with barium hydroxide. The insoluble precipitate which forms under these conditions contains the barium salts of certain not yet identified organic phosphoric acids and it is free from nitrogen. By evaporating the filtrate from the above insoluble barium salts a substance is obtained which is rich in nitrogen and which also contains phosphorus in organic combination.

In the further investigation of this soluble nitrogen-containing substance it was found that its aqueous solution gave an insoluble precipitate with lead acetate. The only other salt which gave any precipitate was copper acetate and then only on warming when a

---

* The work reported in this paper was carried out in the I. Chem. Institut der Universität zu Berlin, Berlin, Germany.

[1] *Jour. Biol. Chem.* **12**: 447 (1912); N. Y. Agric. Exp. Sta. Tech. Bull. 22 (1912), and also the preceding article.

[2] This crude compound had been prepared by precipitating the 0.2 per ct. hydrochloric acid extract of wheat bran with alcohol. The resulting precipitate was then purified by repeatedly precipitating from 0.2 per ct. hydrochloric acid with alcohol until a nearly white product was obtained which was easily soluble in cold water and which gave no precipitate with ammonium molybdate. Concerning its preparation, see the above publications.

bluish-white amorphous precipitate was produced, which dissolved completely on cooling.

The aqueous solution of the substance was therefore treated with lead acetate in excess. The resulting precipitate was filtered, washed, and decomposed with hydrogen sulphide. These operations were repeated several times until a perfectly white lead precipitate was obtained. This was finally decomposed with hydrogen sulphide and the solution concentrated in vacuum until a thick, practically colorless syrup remained. On scratching with a glass rod this immediately crystallized to a white solid mass. The substance was recrystallized from water with addition of alcohol. It was then obtained in beautiful colorless star-shaped aggregates of plates or long prisms. On slowly concentrating its aqueous solution it crystallizes in large colorless prisms with pointed ends, being often arranged in star-shaped bundles. It is, however, so soluble in water that it is more expedient to crystallize it from water with addition of alcohol.

The substance was free from bases, also free from nitrogen and sulphur, but it contained phosphorus in organic combination. Analysis showed that it was inosite monophosphate, $C_6H_{13}O_9P$, or $C_6H_6(OH)_5O \cdot PO(OH)_2$. On cleavage either with dilute sulphuric acid at 120° or higher or with 10 per ct. ammonia at 150° in a sealed tube it decomposes into inosite and phosphoric acid.

Inosite monophosphate has not been known previously, so far as we are aware, and we believe that this is the first time that it has been isolated.

In connection with the " phytin " problem it is interesting to note that a compound like inosite monophosphate exists in nature. Clarke[3] in a recent paper reports the isolation from wild Indian mustards of certain crystalline strychnine salts of what appears to be inosite tetra- and diphosphate in addition to inosite hexaphosphate. It appears probable, therefore, that in certain plants the organic phosphoric acids may be present not only as phytic acid or inosite hexaphosphate, $C_6H_{18}O_{24}P_6$, but also as lower phosphoric acid esters of inosite. From wheat bran, for instance, we have been unable to isolate any inosite hexaphosphate. The insoluble barium salts of the organic phosphorus compound obtained from this material are evidently mixtures of various organic phosphoric acids, either lower inosite phosphates or phosphoric acid esters of other carbohydrates. However, we have been unable, so far, to separate any homogeneous substance from this mixture.

The isolation of inosite monophosphate only succeeded because its properties are so different from those of the other organic phosphoric acids which exist in wheat bran — for instance, its easily soluble barium salt permitted its separation from the other acids which give insoluble barium salts.

[3] *Jour. Chem. Soc.* 105: 535. 1914.

At present we have no data as to the quantitative percentage of inosite monophosphate in wheat bran. We hope, however, to make some determinations in this direction later. We wish to reserve the further study of the physiological properties of this substance in connection with the general investigation which is being carried out at this Station. We also beg to reserve the study of the cleavage products obtained under different conditions and other derivatives of inosite monophosphates.

EXPERIMENTAL PART.

The crude nitrogen-containing substance[4] was dissolved in water and a concentrated solution of lead acetate added in excess. The resulting precipitate was filtered, washed thoroughly in cold water and then suspended in hot water and decomposed with hydrogen sulphide. The lead sulphide was filtered off and the filtrate boiled to expel hydrogen sulphide. The solution was then strongly acid to litmus and it had a sharp acid taste. It was again precipitated as above three times with lead acetate. The pure white colored lead precipitate which was finally obtained was decomposed with hydrogen sulphide. The filtrate was concentrated in vacuum at a temperature of 40°–45° and then dried in vacuum over sulphuric acid until a thick, practically colorless syrup remained. On scratching with a glass rod this immediately began to crystallize, forming a white, solid mass. It was very soluble in water, but insoluble in alcohol. It was extracted several times with 95 per ct. alcohol, filtered and washed in absolute alcohol and ether and allowed to dry in the air. For recrystallization it was dissolved in a small quantity of water and absolute alcohol added until the solution turned slightly cloudy. On scratching, the substance began to crystallize. After standing in the ice chest over night it had separated in large colorless plates or prisms arranged in star-shaped aggregates. It was recrystallized a second time in the same manner.

The substance was free from bases and also free from nitrogen and sulphur, but it contained organically bound phosphorus. The aqueous solution gave no precipitate with ammonium molybdate on being kept at a temperature of 65° C. for some time but after decomposing by the Neumann method it gave an immediate precipitate of ammonium phosphomolybdate with this reagent.

The substance has no sharp melting point. When rapidly heated in a capillary tube it softens at 200° C. and decomposes under effervescence at 201°–202°; when slowly heated it begins to soften at 188° and melts under decomposition at 190°–191° (uncorrected).

It is optically inactive. A 10 per ct. solution in a 1 dcm. tube shows no rotation.

[4] From its isolation from wheat bran see *Journ. Biol. Chem.* 12: 456 (1912); N. Y. Agric. Exp. Sta. Tech. Bull. 22, p. 10 (1912), and also the preceding article.

For analysis .it was dried at 100° in vacuum over phosphorus pentoxide but it did not lose in weight.

0.1550 gram subst. gave 0.0749 gm. H_2O and 0.1566 gm. CO_2.

0.0766 gram subst. gave 0.0325 gm. $Mg_2P_2O_7$.

Found: C = 27.55; H = 5.40; P = 11.82 per ct.

For inosite monophosphate, $C_6H_{13}O_9P = 260$.

Calculated: C = 27.69; H=5.00; P = 11.92 per ct.

Titrated against barium hydroxide, using phenolphthalein as indicator, it forms the neutral barium salt, $C_6H_{11}O_9P$ Ba.

0.1985 gram subst. required 7.6 c.c. $\frac{N}{5}$ Ba(OH)$_2$.

For $C_6H_{11}O_9P$ Ba, calculated: 7.6 c.c. $\frac{N}{5}$ Ba(OH)$_2$.

PROPERTIES OF INOSITE MONOPHOSPHATE.

The acid is very soluble in water. The aqueous solution shows a strong acid reaction to litmus and it has a sharp, somewhat astringent, acid taste. It is insoluble in alcohol, ether and the other usual organic solvents.

Its aqueous solution gives no precipitate with barium hydroxide or with calcium or barium chloride; ammonia produces no precipitate in these solutions but the addition of alcohol causes white amorphous precipitates. Silver nitrate produces no precipitate even in a solution neutralized with ammonia. When alcohol is added to the solution containing silver nitrate a white amorphous precipitate is produced which dissolves on warming; on cooling the silver salt separates in small, round crystal aggregates. It gives no precipitate with ferric chloride or mercuric chloride nor with copper sulphate. In the cold no precipitate is produced with copper acetate, but on warming .this solution a bluish-white precipitate separates which again dissolves completely on cooling.

With excess of lead acetate a white, heavy amorphous precipitate is formed which is but slightly soluble in dilute acetic acid, but readily soluble in dilute hydrochloric or nitric acid. ·Ammonium molybdate produces no precipitate in either dilute or concentrated aqueous solutions.

The acid crystallizes without water of crystallization from either water or dilute alcohol.

The aqueous solution of inosite monophosphate does not precipitate egg albumen, differing in this respect from phytic acid.

CLEAVAGE OF INOSITE MONOPHOSPHATE INTO INOSITE AND PHOSPHORIC ACID.

I. ACID HYDROLYSIS.

The acid, 0.35 gram, was heated in a sealed tube with 15 c.c. of 3 per ct. sulphuric acid to 120°–125° for about $3\frac{1}{2}$ hours. After cooling, the liquid was of a pale straw color. The sulphuric and phosphoric acids were precipitated with barium hydroxide and the excess of barium hydroxide removed with carbon dioxide. The filtrate was evaporated to dryness on the water-bath. The residue gave no precipitate with ammonium molybdate, but after decomposing by the Neumann method, a heavy precipitate of ammonium phosphomolybdate was obtained showing that only a portion of the acid had' been hydrolyzed under the above conditions. The residue, however, contained some inosite which was isolated as follows: The substance was taken up in a few cubic centimeters of hot water, a little more than an equal volume of alcohol was added which caused a voluminous white amorphous precipitate consisting of the barium salt of the unchanged inosite monophosphate. After filtering, the precipitate was again treated with water, again precipitated with alcohol and filtered. The filtrates were evaporated on the water-bath, taken up in a little water and the inosite brought to crystallization by the addition of alcohol and ether. It crystallized in the usual needle-shaped crystals. After standing several hours in the ice chest the crystals were filtered, washed in alcohol and ether and dried in the air. Yield 0.06 gram. It gave the reaction of Scherer and melted at 224° C. (uncorrected).

II. ALKALINE HYDROLYSIS.

Another portion of the acid, 0.4 gram, was heated in a sealed tube with 10 c.c. of 10 per ct. ammonia for six hours to 120°. The solution then contained some free phosphoric acid as it gave a precipitate with ammonium molybdate but the greater portion of the acid remained unchanged. It was found impossible to isolate any inosite from this reaction mixture.

The residue was therefore again heated in a sealed tube with 10 per ct. ammonia for about $4\frac{1}{2}$ hours to 150°. In this case complete hydrolysis had taken place, and after isolating the inosite in the usual way 0.15 gram was obtained. This was recrystallized three times from dilute alcohol with addition of ether and was then obtained in colorless needles free from water of crystallization. It then melted at 224° C. (uncorrected), and it gave the reaction of Scherer. The identity of the substance was further confirmed by the analysis.

0.1206 gram subst. gave 0.0755 gm. H_2O and 0.1761 gm. CO_2.
Found: C = 39.82; H = 6.97 per ct.
For $C_6H_{12}O_6 = 180$.
Calculated: C = 40.00; H = 6.66 per ct.

The author wishes to express his appreciation and thanks to His Excellency Prof. E. Fischer and to Prof. H. Leuchs for the kind interest.which they have shown in the work reported in this paper.

TECHNICAL BULLETIN No. 37.   DECEMBER, 1914.

New York Agricultural Experiment Station.

GENEVA, N. Y.

STUDIES RELATING TO THE CHEMISTRY OF MILK AND CASEIN.

I. THE CAUSE OF ACIDITY OF FRESH MILK OF COWS AND A METHOD FOR THE DETERMINATION OF ACIDITY.
LUCIUS L. VAN SLYKE AND ALFRED W. BOSWORTH.

II. THE PHOSPHORUS CONTENT OF CASEIN.
ALFRED W. BOSWORTH AND LUCIUS L. VAN SLYKE.

III. THE ACTION OF RENNIN ON CASEIN.
(Second paper.)
ALFRED W. BOSWORTH.

PUBLISHED BY THE DEPARTMENT OF AGRICULTURE

STUDIES RELATING TO THE CHEMISTRY OF MILK AND CASEIN.

SUMMARY.

I. The acidity of fresh milk is due to the presence of acid phosphates. Titration of phosphoric acid with alkali, in the presence of calcium salts, results in hydrolysis of dicalcium phosphate formed during the titration, whereby free calcium hydroxide and phosphoric acid are first formed and then calcium hydroxide unites with more dicalcium phosphate to form insoluble tricalcium phosphate. As a result of these reactions more alkali is required to make a solution, containing calcium and phosphoric acid, neutral to phenolphthalein than is required in the absence of calcium. The calcium must be removed previous to titration by treatment of 100 c.c. of milk with 2 c.c. of saturated solution of neutral potassium oxalate.

II. The amount of phosphorus in casein has been commonly given as about 0.85 per ct. By treating a solution of casein in dilute NH_4OH with ammonium oxalate and an excess of NH_4OH and letting stand 12 hours the phosphorus content is reduced to about 0.70 per ct. This lower percentage can not be explained as being due to hydrolysis of casein and splitting off of phosphorus. While some of the casein is hydrolyzed, this portion does not enter into the final preparation and does not affect its composition, because the hydrolyzed portion is not precipitated by acetic acid while the unhydrolyzed part is. The higher figure ordinarily given is due to the presence of inorganic phosphorus (dicalcium phosphate) carried from the milk into the precipitated casein and not entirely removed under the usual conditions of preparation. The lower figure corresponds very closely to two atoms of phosphorus (0.698 per ct.) in the casein molecule. Analyses of various preparations of casein containing varying amounts of ash show a general correspondence between the ash and phosphorus content.

III. The similarity between the composition of casein and paracasein, and the fact that casein has been shown to have a molecular weight of 8888 + and a valency of 8, while paracasein has been shown to have a molecular weight of 4444 + and a valency of 4,[1] seems to be evidence enough for concluding that the transformation of casein into paracasein is a process of hydrolytic splitting, one

---

[1] Van Slyke and Bosworth. N. Y. Agrl. Expt. Sta. Tech. Bull. No. 26, and *Journ. Biol. Chem.*, 14:227.

molecule of casein yielding two molecules of paracasein, and that
this splitting of casein is not accompanied by a cleavage of any of the
elements contained in the original casein molecule.

I. THE CAUSE OF ACIDITY OF FRESH MILK OF COWS AND A METHOD FOR THE DETERMINATION OF ACIDITY.

LUCIUS L. VAN SLYKE AND ALFRED W. BOSWORTH.

INTRODUCTION.

The usual method employed in determining the acidity of milk
is to add a few drops of a solution of phenolphthalein as indicator
to 100 c.c. of milk and then titrate with $\frac{N}{10}$ NaOH. By the use
of this method it is found that 100 c.c. of milk, when strictly fresh,
will require the addition of 15 to 20 c.c. of the alkali in order to
produce a faint but permanent pink coloration.

The acidity of fresh milk has been commonly attributed to the
presence of acid phosphates and casein, and we will now consider
the relation of these constituents to milk acidity.

That the acidity of milk is due to the presence of acid phosphates
(MH_2PO_4) is indicated by the fact that milk is strongly alkaline
to methyl orange. Further, it is well known that phosphates can
not be titrated with any degree of accuracy in the presence of
calcium salts, due to the fact that some of the insoluble dicalcium phosphate ($CaHPO_4$), which is formed during the titration,
hydrolyzes, changing into calcium hydroxide and phosphoric acid,
and then the calcium hydroxide unites with more dicalcium phosphate, forming tricalcium phosphate ($Ca_3P_2O_8$).[2] These facts may
be represented by the following equations:

(1) $CaHPO_4 + 2 H_2O \rightleftarrows Ca(OH)_2 + H_3PO_4$
(2) $2 CaHPO_4 + Ca(OH)_2 \rightarrow Ca_3P_2O_8 + 2 H_2O.$

That tricalcium phosphate is formed during the titration of any
solution containing phosphoric acid and calcium salts is easily
demonstrated by an anlysis of the precipitate always appearing;
this precipitate is tricalcium phosphate, which is characterized
by its appearance, varying from a flocculent to a gelatinous condition according to the concentration of the calcium and phosphates in the solution.

Dibasic phosphates are neutral to phenolphthalein and monophosphates are acid to this indicator; phosphoric acid, therefore,
acts as a diabasic acid to phenolphthalein. In the reaction represented above, we have, in place of the original molecule of neutral
dicalcium phosphate, one molecule of free phosphoric acid, whereby

[2] Cameron and Hurst. *Journ. Amer. Chem. Soc.*, 26: 905. 1904.

the acidity as measured by titration is increased over what it would be if no such reaction occurred. These facts serve to explain some results obtained by us in connection with the study of certain problems relating to milk.

We have found that when we titrate whole milk with alkali, in the usual way and then similarly titrate the serum obtained by filtering the milk through a porous procelain filter, the titration figure given by the whole milk is about double that obtained with the serum. For example, 100 c.c. of whole milk may show an acidity of 17 c.c. of $\frac{N}{10}$ alkali, and 100 c.c. of serum, 8 c.c. This difference has ordinarily been interpreted as being due to the acidity of milk casein, but in a future paper we shall show that casein is present in fresh milk as a calcium caseinate that is neutral to phenolphthalein. The other constituents removed from the milk by filtering through porous porcelain are fat and dicalcium phosphate, both of which are also neutral to phenolphthalein. From the illustration given above, the titration figure of the residue on the filter would appear to be 9 (17−8) for 100 c.c. of milk, though in reality the reaction is neutral. We believe that the cause of this discrepancy is to be found in the dicalcium phosphate which is present in the whole milk but which is not present in the serum. Its presence in the milk permits the formation of relatively large amounts of phosphoric acid and tricalcium phosphate, requiring the use of the increased amounts of $\frac{N}{10}$ alkali (17 c.c.) to neutralize the milk, as compared with the amount (8 c.c.) needed to neutralize the serum. We have been led by such results to believe that the acidity of milk, as usually determined, is about twice what it should be.

The disturbing influence of calcium salts in the presence of phosphates has been studied by Folin[3] in connection with the determination of acidity in urine; he was able largely to overcome the difficulty by the addition of neutral potassium oxalate, by which the calcium is removed in the form of the insoluble oxalate. He showed that by this preliminary treatment correct titration figures could be obtained for monocalcium phosphate which, without such treatment, gives figures that are remote from the calculated acidity.

METHOD FOR DETERMINING THE ACIDITY OF MILK.

Making use of Folin's procedure, and, before titrating with alkali, adding to milk some saturated solution of neutral potassium oxalate, we are able to obtain figures which conform more closely to the results indicated as accurate by other considerations.

The method, as modified by us for the determination of acidity in milk, whether fresh or otherwise, is as follows:

Measure 100 c.c. of milk into a 200 c.c. Erlenmeyer flask, add 50

[3] *Amer. Journ. of Physiol.*, 9:265. 1903.

c.c. of distilled water and 2 c.c. of a saturated solution of neutral potassium oxalate, allow the mixture to stand not less than two minutes and then titrate with $\frac{N}{10}$ NaOH. Since most solid potassium oxalate is acid, care must be taken to prepare a solution that is really neutral, which may be done in the following way: A saturated solution of ordinary potassium oxalate is prepared and decanted from the solid residue. To this solution is added 1 c.c. of phenolphthalein solution and then, drop by drop, enough normal NaOH solution to produce a permanent faintly pink coloration.

In the following table is given the acidity of 21 samples of milk from individual cows, as determined by the two methods, with and without addition of neutral potassium oxalate.

NUMBER OF SAMPLE.	AMOUNT OF $\frac{N}{10}$ NaOH REQUIRED TO NEUTRALIZE 100 C.C. OF MILK.	
	Before addition of neutral potassium oxalate.	After addition of neutral potassium oxalate.
	c.c.	c.c.
1	15	6.4
2	15.2	7.0
3	15.6	6.8
4	16.0	6.8
5	17.0	8.0
6	17.0	8.0
7	17.2	8.0
8	17.6	9.0
9	17.8	8.8
10	18.0	9.0
11	18.2	9.6
12	18.4	9.6
13	18.4	9.4
14	18.6	9.4
15	18.6	9.4
16	19.0	9.4
17	19.2	10.0
18	19.4	10.4
19	20.0	9.8
20	22.0	12.8
21	23.8	14.0

Potassium oxalate is a poisonous substance. If the method as outlined above is used for the determination of the acidity of milk, extreme care should be taken to see that the potassium oxalate solution is kept in a bottle properly labeled and marked with the word POISON.

II. THE PHOSPHORUS CONTENT OF CASEIN.

ALFRED W. BOSWORTH AND LUCIUS L. VAN SLYKE.

INTRODUCTION.

In a previous paper[4] from this laboratory, a method has been described for preparing casein practically ash-free, the last portion of calcium being removed by treating a solution of the casein in dilute NH_4OH with ammonium oxalate and excess of NH_4OH, and then allowing the mixture to stand about twelve hours. Casein thus prepared contains about 0.71 per ct. of phosphorus. The accuracy of this figure has been questioned,[5] because it is considerably lower than that (about 0.85 per ct.) hitherto commonly accepted as correct. The suggestion has been made that the lower figure is due to the splitting off of phosphorus from the casein molecule as the result of hydrolysis caused by prolonged contact with NH_4OH.

It is the purpose of this paper to present the results of an experimental study relating to the effects of partial hydrolysis of casein on the phosphorus content of casein preparations and also to offer an explanation as to why the higher figures that have been usually reported for the percentage of phosphorus in casein are not correct.

In connection with investigations recently carried on in this laboratory, the results of which have not yet been published, certain facts have been developed which appear to explain why the high figure usually accepted for the phosphorus content of casein is inevitably obtained in consequence of the method employed in making casein preparations. Two of the constituents of cow's milk are present in the form of colloidal solution, calcium caseinate and dicalcium phosphate. These two compounds appear to have a strong attraction for each other, as shown by the fact that, when casein is separated from milk by means of either centrifugal force or precipitation with a dilute acid, the casein always carries with it more or less dicalcium phosphate. It is evident, then, that in preparing casein by the usual method in which care is taken to avoid an excess of both acid and alkali, it is practically impossible to remove this phosphate completely. In order, therefore, to ascertain the true phosphorus content of casein, it is obviously necessary that the preparation be free from inorganic phosphorus and this can be accomplished only by removing all of the calcium. Several methods have been tried in this laboratory to effect this, and the one finally found to be the most satisfactory is that described in a previous paper, referred to above.

---

[4] N. Y. Agrl. Expt. Sta. Tech. Bull. No. 26, and *Journ. Biol. Chem.* 14:203. 1913.

[5] Harden and Macallum. *Biochem. Journ.* 8:90.

Further, a good reason for believing that the lower figure more closely approximates the truth than the higher one hitherto commonly accepted as correct is the relation of phosphorus to the molecular weight of casein. In a previous paper[6] it was shown that the molecular weight of casein is approximately 8888. Now, if the casein molecule contains two atoms of phosphorus, the percentage of phosphorus is 0.698, while the phosphorus content would be 1.046 per ct. if there were three atoms of phosphorus. The figure (0.85 per ct.) heretofore regarded as correct represents, therefore, on account of the presence of impurities in the preparation, neither two atoms nor three atoms of phosphorus, while the lower figure (0.71 per ct.) represents almost exactly two atoms.

Coming now to the criticism made that an excess of NH_4OH in contact with casein for twelve hours causes hydrolysis, resulting in the formation of inorganic phosphorus, there is reason to believe that, whatever hydrolysis takes place, it does not necessarily interfere with the composition of the final preparation, because, as will be shown, the products of hydrolysis are not precipitated by dilute acetic acid and therefore form no part of the completed preparation which is pure, unhydrolyzed casein.

EXPERIMENTAL.

After giving the ash and phosphorus content of several preparations of casein, we will present the results of a study of two special preparations of casein which were subjected to varying conditions in order to ascertain whether hydrolysis affects the phosphorus content of casein preparations.

Ash content and phosphorus content of casein.—The percentages of ash and phosphorus in five samples of casein prepared in this laboratory during the past seven or eight years are as follows:

SAMPLE.	Ash.	Phosphorus.
	Per ct.	Per ct.
1	0.06	0.710
2	0.39	0.732
3	0.61	0.830
4	0.61	0.839
5	3.93	0.941

The results show that increase of ash is accompanied by an increase of phosphorus.

[6] N. Y. Agrl. Expt. Sta. Tech. Bull. No. 26, and *Journ. Biol. Chem.*, 14:228. 1913.

Phosphorus content of casein preparations treated in different ways.
In order to study the effect of treating casein in different ways
upon the content of phosphorus, and especially to ascertain what
effect partial hydrolysis may have upon the phosphorus content
of casein preparations, two preparations of casein were made and
each of these was treated in the manner described below.

Preparation A was made in the usual way, treating alternately
with dilute acetic acid and ammonia, avoiding an excess of each
reagent. This preparation contained 0.857 per ct. of phosphorus.

Preparation B was made according to the method given in a
previous paper,[7] the distinctive feature of which is treatment of
a solution of casein in dilute alkali with ammonium oxalate and
excess of alkali. This preparation contained 0.711 per ct. of phosphorus.

(1) Treatment with excess of ammonia. Each of preparations
A and B (20 grams) was dissolved in dilute NH_4OH and an excess
of the same reagent was added; after standing twelve hours at
37° C., the solution was centrifugalized and filtered, the casein
in the filtrate being then precipitated with dilute acetic acid.
This precipitated casein was washed, redissolved, reprecipitated
and finally washed with water, alcohol and ether.

In the case of preparation A, the yield was 14 grams, containing 0.841 per ct. of phosphorus; in the case of preparation B, the
yield was 15 grams and the phosphorus content 0.713 per ct.

The decreased yield in each case was due in part to hydrolysis
of casein and in part to mechanical losses. It is evident that
partial hydrolysis of casein preparations has no effect on the percentage of phosphorus in the unhydrolyzed casein that is recovered.

(2) Treatment with ammonium oxalate and excess of ammonia.
Each of preparations A and B (20 grams) was dissolved in dilute
NH_4OH and then ammonium oxalate and an excess of NH_4OH
added, the mixture being allowed to stand twelve hours at 37° C.
The casein was separated as before.

In the case of preparation A, the yield was 14 grams, containing 0.723 per ct. of phosphorus; in the case of preparation B, the
yield was 14.5 grams, containing 0.71 per ct. of phosphorus.

In these two experiments, hydrolysis of casein by alkali has no
effect upon the percentage of phosphorus in the casein finally recovered. In the case of preparation A, the phosphorus content is
reduced from 0.857 to 0.723 per ct., as a result of the removal of
calcium phosphate from the casein preparation. In the case of
preparation B, the phosphorus content remains the same as in the
original preparation, because the casein used had already been
subjected to treatment with ammonium oxalate and excess of

---

[7] Loc. cit.

NH_4OH, the calcium phosphate having been removed as completely as practicable.

(3) Treatment as in (2) but prolonged. Preparation B (20 grams) was treated as in the preceding experiment, except that the mixture was allowed to stand seventy-two hours (instead of twelve) at 37° C. The amount of casein recovered was 12.4 grams containing 0.721 per ct. of phosphorus. The prolonged treatment, giving opportunity for increased hydrolysis of casein, did not change the percentage of phosphorus in the casein recovered.

III. THE ACTION OF RENNIN ON CASEIN.

(Second paper.)

ALFRED W. BOSWORTH.

INTRODUCTION.

In order to determine if the change from casein to paracasein results in the cleavage of any of the elements contained in the casein molecule it is imperative that pure casein be used as a standard of comparison, and that the rennin activity be positively differentiated from any further proteolytic activity of the enzyme under consideration, for it is quite evident that " Rennin action is probably a hydrolytic cleavage and may be considered the first step in the proteolysis of casein. It would follow from this that the action now attributed to rennin may be produced by any proteolytic enzyme." [8]

EXPERIMENTAL.

Pure casein and paracasein were prepared according to the methods previously published.[9] Pure paracasein was also prepared by allowing trypsin to act upon fat-free milk after the addition of calcium chloride, and the curd produced was purified according to the method referred to. The use of an excess of ammonia as prescribed has been criticised by Harden and Macallum[10] who claim that preparations made in that way may have a low phosphorus content due to the cleavage of phosphorus from the casein molecule by the action of the ammonia. In the preceding paper it has been shown that this criticism does not hold. The analyses of the preparations are given in the table.

---

[8] Bosworth. N. Y. Agrl. Expt. Sta. Tech. Bull. No. 31; also *Journ. Biol. Chem.* 15:236.

[9] Van Slyke and Bosworth. N. Y. Agrl. Expt. Sta. Tech. Bull. No. 26; also *Journ. Biol. Chem.* 14:203.

[10] Harden and Macallum. *Biochem. Journ.* 8:90.

	Casein.	Paracasein by rennin.	Paracasein by trypsin.
Moisture............................	1.09	1.63	1.27
Carbon in dry substance...............	53.50	53.50	53.47
Hydrogen in dry substance.............	7.13	7.26	7.19
Oxygen in dry substance..............	*22.08	*21.94	*22.04
Nitrogen in dry substance.............	15.80	15.80	15.78
Phosphorus in dry sybstance...........	0.71	0.71	0.71
Sulphur in dry substance...............	0.72	0.72	0.72
Ash in dry substance.................	0.06	0.07	0.09

* By difference.

These figures show that the composition of paracasein is the same irrespective of the enzyme used to produce it. The figures also show that casein and paracasein have the same percentage composition, which excludes the possibility that cleavage of any of the elements of casein is a result of its transformation into paracasein by enzymes.

Harden and Macallum, in their paper, conclude that " The conversion of caseinogen into casein by enzyme action is accompanied by the cleavage of nitrogen, phosphorus and calcium.[11] " It seems more probable to us that this cleavage follows rather than accompanies the conversion in question, and is to be attributed to a continuation of proteolytic activity by the enzyme beyond the point where casein has been changed to paracasein. This point was emphasized in my first paper on the action of rennin on casein.[12]

[11] The English caseinogen is equivalent to the American casein. The English casein is equivalent to the American paracasein.

Harden and Macallum give the nitrogen-phosphorus ratio of casein as $N : P = 100 : 5.6$. The high phosphorus content of their casein preparations [0.87 to 0.90 per ct.] would seem to indicate the presence of considerable inorganic phosphorus. If our figures are correct for the nitrogen and phosphorus content of casein [15.80 per ct. N, 0.71 per ct. P] the ratio would be $N : P = 100 : 4.50$. In only one of their experiments conducted to show the loss of phosphorus from the casein molecule was the N-P ratio reduced to 4.50.

[12] Bosworth. N. Y. Agrl. Expt. Sta. Tech. Bull. No. 31; also *Journ. Biol. Chem.* 15:231.

8 g b

TECHNICAL BULLETIN No. 38.   NOVEMBER, 1914.

New York Agricultural Experiment Station.

GENEVA, N. Y.

CULTURE MEDIA FOR USE IN THE PLATE METHOD OF COUNTING SOIL BACTERIA.

H. JOEL CONN.

PUBLISHED BY THE DEPARTMENT OF AGRICULTURE

CULTURE MEDIA FOR USE IN THE PLATE METHOD OF COUNTING SOIL BACTERIA.

H. JOEL CONN.

SUMMARY.

1. Two new culture media have been tested to determine their merits when employed in the plate method of counting soil bacteria. One is a soil-extract gelatin, the other an agar medium containing no organic matter except the agar, dextrose and sodium asparaginate.

2. The soil-extract gelatin is recommended primarily for use when the plate method is employed as a preliminary procedure in a qualitative study of soil bacteria. Its advantages are that the colonies produced upon it by different types of bacteria are fairly distinct in appearance, and rather more of the soil bacteria produce colonies upon it than upon any other medium investigated. Its disadvantages are such that they do not render it less satisfactory for qualitative purposes although they might make its use inadvisable in quantitative work.

3. The chief advantage of the asparaginate agar is that it contains no substance of indefinite composition except the agar itself. This ought to allow comparable results to be obtained by its use, even though the work be done by different men and in different laboratories. It is therefore especially adapted to quantitative work.

4. Four other media have been compared with these. They are those recommended for use in soil bacteriological studies by Fischer, by Lipman and Brown, by Temple, and by Brown. For qualitative purposes they are all distinctly inferior to gelatin. For quantitative work they are undesirable because they contain substances of indefinite chemical composition. It has been found that no one of the five agar media has a distinct advantage over any of the others in the matter of the total counts obtained by their use.

INTRODUCTION.

Ever since the bacteriology of soil was first studied, one of the most common lines of investigation has been to determine the number of bacteria living in different soils. It was thought at first that the number of bacteria present was proportional to the productivity of the soil. Investigation, however, soon proved that the rule held only in a very general way, and that exceptions were extremely numerous. Quantitative work, therefore, fell more and more into disfavor until in 1902 Remy[1] stated that mere determinations of the number of bacteria were of no use. Remy realized the importance of knowing not only the numbers of bacteria, but also the kinds present and their functions. He considered a complete study of this sort, however, too colossal a task to be undertaken. As a practical substitute for a complete qualitative study, Remy suggested a method of obtaining a qualitative knowledge of soil bacteria without counting them or separating the different kinds from each other. This was accomplished by measuring the chemical changes which the total flora of any soil was capable of producing when inoculated into special liquid media. Bacteriological methods, today, have improved to such an extent that it may soon be possible to make a more complete study of soil bacteria, including determinations of the number of each kind of bacteria present as well as the total number, and also of the functions of the various bacteria. Quantitative as well as qualitative methods, however, must be perfected before a complete study of this sort becomes possible.

The usual method employed in quantitative work has been to count the colonies developing upon a plate of nutrient gelatin or agar inoculated with a small amount of soil infusion of definitely known dilution. It is known that the composition of the nutrient gelatin or agar has considerable influence upon the results, but the best possible composition has not yet been determined. The present investigation is a study of the relative merits of various culture media for this purpose. The results are to be considered as merely preliminary; but they are published as an aid to others who are striving after satisfactory media for soil bacteriological work.

USES AND LIMITATIONS OF THE PLATE METHOD.

The weaknesses of the plate method of counting bacteria are too well known to need much discussion. It is an indirect method, for by its use the bacteria are counted only by means of the colonies they produce on the plates. In interpreting the colony count as though it were a count of the bacteria themselves, it has to be assumed that every bacterium mixed with the culture medium develops into a macroscopic colony, an assumption that is not justified unless the composition of the medium, the temperature of incubation, and the length of time allowed to elapse before counting are such as to permit the growth of every kind of organism present, and unless the bacteria are so well separated from each other that no colony represents more than a single individual. These conditions have never been fully met, and probably never can be.

---

[1] Remy, Th. Bodenbakteriologische Studien. *Centbl. Bakt.*, **Abt. II,** 8:657–662, 699–705, 728–735, 761–769. 1902.

A still greater weakness results from the fact that the culture media in general use are such that there is no means of being sure in any case to what extent these conditions have been met. The fault of the ordinary culture media is that they contain materials of indefinite composition, different lots of which undoubtedly vary sufficiently in composition so that conditions are at times more favorable for bacterial growth than at others. The result is that the counts obtained by different workers, or by the same worker when using different batches of media, may vary greatly, even though there be no variation in the actual number of bacteria present.

The fact that the plate method gives incomplete counts when applied to soil has been illustrated by comparing it with the only other method that has been proposed for counting soil bacteria. Hiltner and Störmer[2] suggested that liquid instead of solid media be used for making quantitative determinations. They recommended the use of four different liquid media in making each test, each medium adapted to the growth of some particular group of soil bacteria. Their method was to inoculate each medium with small portions of soil infusion of many different dilutions, some of them so dilute as to cause no reaction to take place in the medium into which they were introduced. Having determined how great a dilution was necessary in inoculating each medium before tubes could be obtained in which bacteria adapted to that medium were lacking, a simple calculation sufficed to show the approximate number of each of these groups present in the soil investigated. This dilution method, according to Löhnis,[3] gives higher counts than the plate method, a fact which suggests that many bacteria are overlooked when the latter method is used.

In spite of this well-known weakness of the plate method, it is in common use today, while Hiltner and Störmer's method is rarely employed. This is partially to be explained by the relative convenience of the two methods. Hiltner and Störmer's method is cumbersome, while poured plates furnish a simple and convenient means of testing several samples in a comparatively short time. A second advantage of the plate method is that it is possible to isolate pure cultures of the various bacteria from the colonies that develop on the plates. The study of these pure cultures gives a more thorough qualitative knowledge of the soil flora than can be obtained by Hiltner and Störmer's method.

This second advantage of the plate method must be made even greater before the procedure becomes of the greatest possible value

---

[2] Hiltner, L., and Störmer, K. Studien über die Bakterienflora des Ackerbodens, mit besonderer Berücksichtigung ihres Verhaltens nach einer Behandlung mit Schwefelkohlenstoff und nach Brache. *Kaiserliches Gesundheitsamt. Biol. Abt. Land. u. Forstw.*, 3:445–545, 1903.

[3] Löhnis, F. Zur. Methodik der bakteriologischen Bodenuntersuchung II. *Centbl. Bakt. Abt. II*, 14:1–9, 1905.

in qualitative work. Qualitative bacteriological analysis depends upon the ease with which the colonies of different types of bacteria can be distinguished from each other. The differences in appearance of the colonies result from certain peculiarities possessed by the bacteria themselves, such as their nutritive requirements or methods of growth. The extent to which these peculiarities are impressed upon the colonies depends upon the conditions of growth furnished by the medium. Unfortunately many of the media in use for the plate method are not favorable to the development of these differences in appearance.

CHARACTERISTICS OF A SATISFACTORY CULTURE MEDIUM FOR THE PLATE METHOD

These limitations of the plate method cannot be entirely overcome; but the technique may be made more serviceable by using a more satisfactory medium. There are at least three important requirements that must be met by any medium before it can be considered perfectly satisfactory for soil work: (1) It must allow the growth of the greatest possible number of soil bacteria. (2) The colonies produced upon it by different types of bacteria must be as distinct as possible in appearance. This requirement, however, need not be met if mere quantitative results are desired. (3) It must be what bacteriologists often term a "synthetic" medium; i. e., of definite chemical composition. This requirement applies especially to quantitative work.

As the plate method serves at least two distinctly different purposes, it may be possible to use two different media, neither of which meets all three requirements. One medium, designed primarily for qualitative work, should fulfill the first and second requirements; the other, intended for quantitative purposes only, should fulfill the first and third.

In the present investigation an agar medium and a gelatin medium have been studied. Both have been tested as to their ability to meet the first of these three requirements. The former has been tested because, like all other gelatin media, it allows good distinctions between the colonies of many kinds of bacteria, and thus fulfills, in part at least, the second requirement. The latter was tested because it contains no material of indefinite chemical composition except the agar itself, and thus nearly fulfills the third requirement.

REVIEW OF LITERATURE.

Previous work along this line has had but one main object in view — that of obtaining a medium allowing the greatest possible number of soil bacteria to produce colonies. The other two requirements just mentioned have been largely overlooked. Some investigators have used the ordinary media of general bacteriological work, such as beef-extract-peptone gelatin, as used by Hiltner and Störmer,[4] or Heyden agar as used by Engberding.[5] None of these were very satisfactory and it was soon concluded that special media must be used in soil work. The simplest modification (e. g. that of Hoffman[6]) differed from the ordinary beef-extract-peptone formulæ only in the substitution of soil-extract for pure water. A slight improvement was claimed for this modified formula, but it has not been generally regarded as sufficient to warrant its continued use.

The recent modifications have all been of a different sort. The best results have been obtained on media low in organic matter. The low organic content of these media undoubtedly holds in check certain rapidly growing organisms that would otherwise prevent the growth of the more numerous but more slowly growing bacteria. Fisher[7] in 1909 described several media of this nature. Early in the following year[8] Fischer recommended another medium, still simpler in composition, which allowed even more soil bacteria to produce colonies. This last medium was an agar to which nothing was added but soil-extract (prepared by extracting with a 0.1 per ct. solution of Na_2CO_3) and potassium phosphate. The advantage of reducing the amount of organic matter was discovered contemporaneously by Lipman and Brown[9] who recommended an agar which contained no nitrogen beyond that furnished in 0.05 gram of peptone per litre. In 1911 Temple[10] recommended a culture medium for soil work which was also low in organic content, although it contained one gram of peptone per litre. Temple states that he could obtain better results with this medium than with Lipman and Brown's formula. In 1913 Brown[11] published a modification of Lipman and Brown's formula, replacing the .05 gram of peptone with one gram of albumin. (For the complete formulæ of the last four media see Table I.) Brown gives the results of six comparative tests that

[4] See footnote 2.

[5] Engberding, D. Vergleichende Untersuchungen über die Bakterienzahl im Ackerboden in ihrer Abhängigkeit von äusseren Einflusscn. *Centbl. Bakt.* Abt. II, 23:569–642, 1909.

[6] Hoffman, C. Relation of Soil Bacteria to Nitrogenous Decomposition. Wis. Agr. Exp. Sta., 23 Ann. Rpt., pp. 120–134, 1906.

[7] Fischer, H. Bakteriologisch-chemische Untersuchungen. Bakteriologischer Teil. *Landw. Jahrb.* 38:355–364, 1909.

[8] Fischer, H. Zur Methodik der Bakterienzählung. *Cenbtl. Bakt.*, Abt. II, 25:457–459,1910. Although similar soil-extract media have been used by other bacteriologists, the directions given by Fischer for preparing this medium are so explicit that it is denoted in the present publication as Fischer's soil-extract agar.

[9] Lipman, J. G., and Brown, P. E. Media for the Quantitative Estimation of Soil Bacteria. *Centbl. Bakt.*, Abt. II, 25:447–454, 1910.

[10] Temple, J. C. The Influence of Stall Manure upon the Bacterial Flora of Soil. *Centbl. Bakt.*, Abt. II, 34:206–223, 1911. Also Ga. Agr. Exp. Sta., Bul. 95:1–34, 1911. (See p. 9 of the latter reference.)

[11] Brown, P. E. Media for the Quantitative Determination of Bacteria in Soils. *Centbl. Bakt.* Abt. II, 38:497–506, 1913. Also Ia. Agr. Exp. Sta. Research Bul. 11:396–407, 1913.

show a slight superiority of this albumin agar both over Lipman and Brown's earlier medium and over Temple's agar.[12] Each of these media has been recommended by a different investigator, and, beyond the tests made by Brown, no comparison between them seems to have been published.

The little weight which the authors of these media have attached to the matter of distinctions in appearance between different kinds of colonies is shown by the fact that they have scarcely ever recommended the use of gelatin media. A few investigators, it is true, such as Hiltner and Störmer[13] used the ordinary beef-extract-peptone gelatin. Hoffman[14] similarly used a beef-extract-peptone gelatin with soil-extract. Fischer[15] mentioned rather casually the use of a gelatin medium containing nothing but gelatin, soil-extract, and 1 per ct. of dextrose, but does not recommend its use on account of the low and irregular counts obtained upon it. A modification of Fischer's formula (containing 0.1 per ct. instead of 1 per ct. dextrose) has been used by the writer with very good results. The use of this last mentioned formula has been referred to already in several publications;[16] but no counts have yet been given to show how this medium compares with others. It is the gelatin medium to which particular attention is to be given in the present paper.

The use of soil-extract, peptone or some other substance of unknown chemical formula in all of these media shows how little stress has been laid upon the importance of having culture media of definite chemical composition. Lipman and Brown's medium alone is fairly satisfactory in this respect. Its authors, indeed, speak of it as a " synthetic " agar, in spite of the fact that it con-

[13] The following is a summary of the six tests, taken from Brown's tables (loc. cit.):

Test	Lipman and Brown's agar.	Brown's albumin agar.	Temple's peptone agar.
1	5,478,000	6,735,000	5,791,000
2	5,200,000	7,775,000	5,225,000
3	4,866,000	7,113,000	5,066,000
4	4,688,000	6,466,000	4,710,000
5	4,560,000	5,999,000	
6	3,086,000	4,158,000	

[13] See footnote 2.
[14] See footnote 6.
[15] See footnote 7.
[16] Conn, H. J. Bacteria in Frozen Soil. *Centbl. Bakt.*, **Abt. II**, 28:422–434, 1910. Bacteria of Frozen Soil, II. *Centbl. Bakt.*, **Abt. II**, 32:70–97, 1911. A Classification of the Bacteria in Two Soil Plats of Unequal Productivity. Cornell Univ. Agr. Exp. Sta., Bul. 338:65–115, 1913. Bacteria of Frozen Soil. N. Y. Agr. Exp. Sta., Tech. Bul. 35:1–20, 1914.

tains agar and peptone. The term "synthetic" is justified only on the assumption that variations in these materials do not affect the growth of bacteria; but Brown states[17] that the form of nitrogen, even when used in such minute quantities, is of great importance in determining the value of the medium. The agar medium tested out in the present work, concerning which a preliminary note has already been published[18], is furnished no organic nitrogen except that contained in sodium asparaginate, and is apparently the first medium ever used for making poured plates from soil that does not contain nitrogen in the form of some indefinite chemical compound.

Table I gives the formulæ of the media preferred respectively by Fischer, by Lipman and Brown, by Temple, and by Brown in the publications already mentioned. It also gives the formulæ of the soil-extract gelatin and asparaginate agar recommended by the writer. The present work is an investigation of these six media.

TABLE I.— COMPOSITION OF VARIOUS CULTURE MEDIA FOR SOIL BACTERIOLOGICAL WORK

CONSTITUENTS.	Fischer's soil-extract agar.	Lipman and Brown's "Synthetic" agar.	Brown's albumin agar.	Temple's peptone agar.	Soil-extract gelatin.	Aspar-aginate agar
Distilled water.....	....	1,000	1,000	....	900	1,000
Tap-water.........	....	....	......	1,000	....	....
Soil-extract........	*1,000	....	......	....	†100	....
Agar..............	12	20	15	15	....	12
Gelatin............	....	....	......	......	120	....
Peptone...........	....	0.05	......	1	....	....
Albumin...........	....	....	1	....	....	....
Sodium asparaginate	....	....	......	....	....	1
Dextrose...........	....	10	10	....	1	1
MgSO₄	....	0.2	0.2	....	....	0.2
K₂H PO₄...........	2	0.5	0.5	....	....	....
NH₄H₂PO₄.........	....	....	......	....	....	1.5
CaCl₂	....	....	......	....	....	0.1
KCl	....	....	......	....	....	0.1
FeCl₃	....	....	......	....	....	Trace
Fe₂(SO₄)₃	....	....	Trace	....	....	....

* Prepared by heating soil for half an hour at 15 pounds pressure with an equal weight of a 0.1 per ct. solution of Na_2CO_3.

† Prepared by boiling soil half an hour with an equal weight of distilled water (see p. 11).

[17] See footnote 11. Ia. Bul. p. 397. *Centbl.* p. 498.
[18] Conn, H. J. A New Medium for the Quantitative Determination of Bacteria in Soil. *Science*, N. S., 39:764, 1914.

GENERAL TECHNIQUE

The technique used by various soil bacteriologists has differed not only in the kinds of culture media used but also in the length of incubation and the temperature employed. The two latter factors are fully as important as the composition of the medium in determining the final count. Fischer incubated his plates at 16-20° C. and counted during the second week. Temple used a temperature of 25° C. for six days. Lipman and Brown held their plates three days at " about 25° C." This last method is open to criticism, for it has been found in the studies reported upon in this bulletin that so many new colonies continue to appear on Lipman's and Brown's media after three days that the count may be as much as three times as high on the tenth day as on the third. The counts published by Lipman and by Brown, indeed, are considerably lower than those obtained upon their media in the course of the present investigation, in which the longer incubation time has been used.

In the present work gelatin plates have been incubated for seven days, agar plates for fourteen. It would undoubtedly have been more satisfactory to hold gelatin plates a few days longer; but as liquefaction often prevents a count under these circumstances, seven days has been chosen for the routine incubation time. In the case of agar plates, on the other hand, very few new colonies develop after the tenth day, and the longer period of incubation seems to be unnecessary. The use of the fourteen-day period was begun before this fact was known, and it was continued throughout the work in order to make all the results comparable.

The temperature used for incubation has been 18° C. The incubator employed[19] is one that can be kept at a very constant temperature; and it has never reached a temperature as high as 19° except on the hottest summer days. In the case of gelatin, the use of this low temperature is very important, because it prevents rapid liquefaction.

The soils chosen for making these tests have been of as great a variety as could be obtained in this locality. They vary in texture from muck to sand. They are of the following various origins: glacial lake deposit (Dunkirk series), glacial till of the New York drumlin area (Ontario series), glacial till from Devonian shales and sandstones (Volusia silt loam), alluvial (Genesee soils) and a limestone residual soil mixed somewhat with glacial materials (Honeoye stony loam). The nomenclature used is that adopted by the Bureau of Soils of the United States Department of Agriculture.[20]

[19] Conn, H. J., and Harding, H. A. An Efficient Electrical Incubator. N. Y. Agr. Exp. Sta., Tech. Bul. 29:1-16, 1913.
[20] U. S. Dept. Agr. Bureau of Soils, Bul. 96, pp. 1-791, 1913. See also Soil Survey of Ontario County, New York, published by this Bureau, 1912; pp. 1-55.

Each sample of soil has been plated in two different dilutions. These dilutions have varied somewhat with the different soils used; but in each test listed in this bulletin, the figures for the different media have invariably been obtained from plates of the same dilution. The dilution chosen for counting has usually permitted about one hundred colonies to develop per plate. Plates have always been made in triplicate. The counts given in the tables represent the average of the three plates, except in cases where one of the three has been lost by liquefaction or otherwise. In any case where only one of the three triplicate plates has given a reliable count, the results have been discarded or else the figures in the table have been marked doubtful.

THE SOIL-EXTRACT GELATIN.

DESCRIPTION OF THE MEDIUM.

The soil-extract gelatin has been used in routine work by the writer for five years. When first used, it was thought to be merely a makeshift that would quickly be superseded by other media upon which soil bacteria could grow more readily, but it has proved so satisfactory that it has been kept in routine use in this laboratory up to the present time, and has been included in such a large majority of the tests reported in this bulletin that it serves as a basis of comparison between the media that are not compared together directly.[21] As already mentioned it is to be recommended for qualitative work because it meets the requirement of showing distinctions in appearance between the colonies of different kinds of bacteria. The data published in this bulletin will show whether it is also adapted to the growth of as large a number of soil bacteria as are the other culture media that have been proposed for soil work.

The preparation of this gelatin is as follows: Soil, heated in an autoclave for an hour at 20 to 25 pounds pressure, is extracted by mixing with an equal weight of distilled water, allowing the mixture to stand cold for twelve hours and then boiling half an hour, restoring the water lost by evaporation, and filtering. In making up each batch of the medium, the soil-extract is diluted with distilled water to one-tenth its natural strength and used for dissolving the gelatin. (Gold Label Gelatin has always been used.) It is probably unnecessary to carry out in detail the whole of this procedure for obtaining soil-extract, but it was followed carefully throughout the present work in the hope that the composition of the soil-extract might be more nearly constant than it would be if the method of preparation were allowed to vary. After dissolving the gelatin in the diluted soil-extract, the medium is clarified by the use of the white

---

[21] The only caution necessary in employing this gelatin is the use of an incubation temperature as low as 18° C. See p. 28.

of egg, as generally recommended for ordinary bacteriological media.
Dextrose is added just before tubing. The reaction is adjusted
to 0.5 per ct. normal acid to phenolphthalein. The formula is given
in the fifth column of Table I (p. 9).

VARIATIONS BETWEEN DIFFERENT BATCHES OF THE MEDIUM.

A strong objection to this gelatin is its indefinite chemical composition. Variations in the composition might easily cause irregularities in the counts. As a matter of fact, however, there has
seldom been any evidence of such variation; but in Table II is
given one instance where it was noticeable. This table shows the
counts obtained in a series of nine platings upon three batches of
soil-extract gelatin all made up from the same lot of soil-extract,
although from different packages of gelatin. Batch I was two and
a half months old at the time of use; batches II and III were made
up fresh, but batch II had been left in a warm room over night
before sterilization and a faintly noticeable decomposition had taken
place. It will be seen that the counts on batch II are considerably
lower than those on batch I. Batch III was used only six times.
Generally it gave a count intermediate between batches I and II.

In this particular case the cause of the poor results from batch II
was undoubtedly its decomposition; but the decomposition was
so very slight that if it had not happened to be accompanied by gas
formation, it might have been overlooked. A similar accident
might easily occur in making up any batch without being noticed.
There are many other opportunities for such variation in composition
of the media. Agar is as liable to these variations as gelatin; again
and again some batch of agar under investigation, apparently made
up in exactly the same manner as the others, has proved unusually
satisfactory or else unusually unsatisfactory. These irregularities,
indeed, are great enough to make the result of any comparison
between two media unreliable unless more than one batch of each
medium has been used.

This fact does not seem to have been fully realized by some
investigators. The six comparative tests given by Brown[22] were
presumably made with a single batch of each medium investigated,
although the author makes no statement to that effect. A glance
at his figures makes it plain that the variations between the counts
he obtained upon the different media are less than those shown in
Table II as occurring between two different batches of gelatin,
and likewise less than those known to occur between different lots
of agar media.

[22] See footnote 12.

TABLE II.— TESTS COMPARING DIFFERENT BATCHES OF SOIL-EXTRACT GELATIN.

Test No.	Date.	Soil type.	BACTERIA PER GRAM DRY SOIL, AS DETERMINED WITH —		
			Batch I — 2½ months old.	Batch II — fresh, but decomposed.	Batch III — fresh; good.
	1914.				
1	Sept. 1	Dunkirk silty clay loam..	24,000,000	10,000,000	....
2	Sept. 2	Dunkirk silty clay loam..	14,000,000	9,000,000	....
3	Sept. 2	Dunkirk silty clay loam..	12,000,000	8,500,000	....
4	Sept. 5	Dunkirk silty clay loam..	35,000,000	23,000,000	....
5	Sept. 5	Dunkirk silty clay loam..	24,000,000	13,000,000	19,000,000
6	Sept. 10	Dunkirk silty clay loam..	21,500,000	13,500,000	*20,000,000
7	Sept. 10	Dunkirk silty clay loam..	39,000,000	25,000,000	32,000,000
8	Sept. 10	Dunkirk silty clay loam..	20,000,000	14,500,000	16,500,000
9	Sept. 11	Ontario fine sandy loam..	....	9,500,000	*9,500,000
10	Sept. 11	Ontario fine sandy loam..	....	9,500,000	11,500,000

* These counts are inexact because of rapid liquefaction.

SIMPLIFICATION OF THE FORMULA OF THE SOIL-EXTRACT GELATIN.

The opportunity for such variations in composition seems, *a priori*, to be greater in the case of this gelatin than with any of the agar media discussed in this paper. Soil-extract is unquestionably of variable composition. Gelatin itself also may be the cause of considerable irregularity. It is more complex in chemical composition than agar and presumably more variable. It may perhaps contain fewer impurities; but it is used in ten times as large quantities as agar, which must result in the introduction of large amounts of whatever impurities it does contain. Lastly, gelatin is a food for many bacteria, and for that reason variations in its composition must have more influence upon bacterial growth than those in agar, which is not ordinarily of nutrient value for bacteria.

In the hope of eliminating some of these causes of variation, an attempt was made, toward the close of the present investigation, to simplify the formula of the gelatin. If any way of purifying the gelatin itself had been known, that would have been undertaken. In the lack of such knowledge, attention was turned to the soil-extract. Eliminating the soil-extract could not prevent the sort of variations shown in Table II, but it might prevent others equally great.

The soil-extract was first replaced by tap-water. The results were so surprisingly successful that both the tap-water and the dextrose were finally eliminated, leaving only a solution of gelatin in distilled water, clarified with white of egg. The results are given in Table III. It will be seen, first, that the tap-water gelatin with

TABLE III.— TESTS OF SIMPLIFIED GELATIN FORMULAE.

Test No.	Date.	Soil Type.	Bacteria Per Gram Dry Soil, as Determined with —					
			Gelatin, soil-extract, dextrose.	Gelatin, tap-water, dextrose.	Gelatin, tap-water.	Gelatin, distilled water, dextrose.	Gelatin, distilled water.	Gelatin, distilled water (unclarified).
	1914.							
1	Mar. 14	Dunkirk fine sand	8,000,000	*11,500,000				
2	Mar. 14	Dunkirk fine sand	26,000,000	26,000,000				
3	Mar. 17	Volusia silt loam	14,000,000	15,000,000				
4	April 3	Dunkirk silty clay loam	31,000,000	30,000,000		33,000,000	30,000,000	
5	April 3	Dunkirk silty clay loam	33,000,000	37,000,000		32,500,000	32,500,000	
6	April 15	Dunkirk silty clay loam	21,500,000				22,000,000	
7	April 15	Dunkirk silty clay loam	32,000,000				28,000,000	
8	April 17	Volusia silt loam	10,000,000	8,000,000		9,000,000	9,000,000	
9	April 17	Volusia silt loam	10,000,000	9,000,000		9,000,000		
10	April 17	Volusia silt loam	21,000,000			14,000,000	14,000,000	
11	April 29	Dunkirk silty clay loam	16,000,000	25,000,000	16,000,000			18,000,000
12	April 29	Dunkirk silty clay loam	15,000,000	19,000,000				
13	May 2	Dunkirk silty clay loam	11,000,000					
14	May 2	Dunkirk silty clay loam	31,000,000				17,000,000	16,000,000
15	May 22	Honeoye stony loam	12,000,000				32,000,000	19,000,000
16	May 22	Honeoye stony loam	16,000,000				12,500,000	10,000,000
17	May 29	Dunkirk silty clay loam	20,000,000	18,000,000	16,000,000	12,000,000	12,000,000	
18	June 4	(A soil from Colorado)	35,000,000	36,000,000	33,000,000			
19	June 4	(A soil from Colorado)	9,000,000	10,000,000	10,000,000			
20	June 5	Norfolk sand	2,000,000		1,200,000			
21	June 5	(A soil from No. Dakota)	6,500,000	6,000,000		5,800,000	1,000,000	
22	Aug. 5	Volusia silt loam	9,500,000		10,600,000		10,300,000	
23	Aug. 5	Volusia silt loam	12,500,000		12,000,000		10,000,000	
24	Aug. 6	Volusia silt loam	8,300,000		8,000,000		6,700,000	
25	Aug. 6	Volusia silt loam	†13,000,000		8,800,000		7,200,000	

No.	Date	Soil				
26	Aug. 6	Volusia silt loam	9,500,000	.....	6,000,000	6,300,000
27	Aug. 8	Ontario loam	†22,000,000	.....	†19,000,000	21,000,000
28	Aug. 8	Ontario loam	25,000,000	.....	27,500,000	22,800,000
29	Aug. 10	Volusia silt loam	†5,500,000	.....	4,500,000	3,600,000
30	Aug. 10	Volusia silt loam	8,500,000	.....	5,800,000	7,000,000
31	Aug. 11	Volusia silt loam	10,500,000	.....	7,700,000	8,000,000
32	Aug. 11	Volusia silt loam	13,500,000	.....	9,600,000	7,200,000
33	Aug. 19	Volusia silt loam	23,500,000	.....	†22,000,000	23,000,000
34	Aug. 19	Dunkirk silty clay loam	29,000,000	.....	22,000,000	‡28,000,000
35	Sept. 1	Dunkirk silty clay loam	24,000,000	16,000,000	18,000,000	.....
36	Sept. 2	Dunkirk silty clay loam	14,000,000	11,000,000	10,500,000	.....
37	Sept. 2	Dunkirk silty clay loam	12,000,000	**13,000,000**	8,800,000	.....
38	Sept. 5	Dunkirk silty clay loam	35,000,000	25,000,000	25,000,000	.....
39	Sept. 5	Dunkirk silty clay loam	24,000,000	17,500,000	16,000,000	.....
40	Sept. 10	Dunkirk silty clay loam	20,000,000	16,500,000	16,500,000	.....
41	Sept. 10	Dunkirk silty clay loam	32,000,000	28,500,000	28,000,000	.....
42	Sept. 10	Dunkirk silty clay loam	16,500,000	14,500,000	17,000,000	.....
43	Oct. 23	Dunkirk silty clay loam	†22,000,000	21,000,000	.....	.....
44	Oct. 23	Dunkirk silty clay loam	30,000,000	†**30,000,000**	.....	.....
45	Oct. 23	Dunkirk silty clay loam	29,000,000	**32,000,000**	.....	.....
46	Oct. 28	Dunkirk silty clay loam	6,200,000	**6,800,000**	.....	.....
47	Nov. 5	Dunkirk silty clay loam	18,000,000	17,300,000	.....	.....
48	Nov. 5	Dunkirk silty clay loam	22,000,000	20,000,000	.....	.....
49	Nov. 6	Dunkirk silty clay loam	22,000,000	‡20,000,000	.....	.....
50	Nov. 6	Dunkirk silty clay loam	16,000,000	**19,000,000**	.....	.....

* Those counts on the simplified formulae that are higher than the corresponding counts on soil-extract gelatin are printed in bold-faced type.

† These counts are inexact because of rapid liquefaction.

‡ In these cases there was such irregularity between the number of colonies upon the parallel plates that satisfactory averages could not be made.

dextrose has given a higher count than the soil-extract gelatin quite often in the earlier tests but only four times in the last sixteen. As different batches of the media were used in the earlier and later tests, it is quite possible that the variation in the counts may have arisen from this cause alone. Secondly it will be seen that the tap-water gelatin without dextrose has rarely given a count as high as that on soil-extract gelatin; but that all the differences are too slight to show an actual advantage for either formula. Thirdly, it will be noticed that the counts on distilled water gelatin with or without dextrose are still more rarely equal to those on the soil-extract gelatin. In this case also the differences are so slight that their significance is doubtful. These tests cannot be construed as showing any reason for using soil-extract rather than tap-water or even distilled water. If further tests show similar results one of the simpler formulæ will unquestionably be considered superior for routine work.

These tests show that either the gelatin, itself, or the white of egg used in clarification has furnished the bacteria with sufficient nutrient matter to cause large numbers of them to develop into colonies. To determine which of these sources was the more important a solution of gelatin was made in distilled water and then used without clarification. The counts obtained on it are given in the last column of Table III. Only four tests of this medium were made; but in two of them the count was higher and in one other almost as high as on the clarified soil-extract gelatin. In spite of the small number of tests made, it seems safe to conclude that gelatin is in itself a very satisfactory culture medium for soil-bacteria.

The use of the soil-extract gelatin was continued, however, throughout the present investigation even though one of the simpler formulæ might have given as good results. Its employment in the earlier work made its continued use valuable as a basis of comparison, and it is plain that the simpler formulæ do not give any better results. A further discussion of the merits of this gelatin follows (pp. 25 to 32) in connection with the discussion of the tables in which it is compared with various agar media.

THE ASPARAGINATE AGAR.

DESCRIPTION OF THE MEDIUM.

The asparaginate agar is intended primarily for quantitative work, as it contains no substance of indefinite chemical composition except the agar itself; but it does not allow such great differences in the appearance of the different colonies as does gelatin. The comparative tests which follow (pp. 25 to 32) will show whether it meets the other important requirement of a medium for quantitative work, that of allowing the growth of the greatest possible number of soil bacteria.

The sole form of organic nitrogen in this agar medium is sodium asparaginate. The formula is given in Table I (p. 9). It is much like the formulæ recommended by Lipman and by Brown, its principal differences being that nitrogen is furnished·in the form of definite chemical compounds only (sodium asparaginate and ammonia phosphate), that it contains only 0.1 per ct. instead of 1 per ct. dextrose, and that it contains the ions Ca and Cl which Lipman and Brown do not use.

In the preparation of the asparaginate agar, the dextrose and sodium asparaginate have been added just before sterilization, so as to avoid any possible effects of the preliminary heating on these substances. The reaction has always been carefully adjusted; because if the acidity is as high as 1.5 per ct. normal (using phenolphthalein as an indicator) the count is appreciably lowered (see Table VIII, p. 25). If it is as low as 0.5 per ct. normal, there is danger of decomposing the ammoniun phosphate and losing the ammonia. The reaction should be between 0.8 per ct. and 1.0 per ct. normal acid to phenolphthalein.[23]

Considerable difficulty has been experienced in clarifying this medium by the ordinary procedure, using the white of egg. Sufficient clarification can be accomplished, however, by heating the medium half an hour at 15 pounds steam pressure in such a way as not to disturb the sediment, and then decanting through a cotton filter. This method of clarification is simpler and is really preferable to the use of white of egg, as it does not introduce into the medium any material of indefinite composition.

TESTS TO DETERMINE THE MOST SATISFACTORY FORMULA.

The exact formula given for this agar in Table I is not to be considered as the only satisfactory combination possible. Tables IV to VIII show the results of a few tests bearing on this point. The conclusions that may be drawn from these tables are somewhat limited by the small number of tests made. The same irregularity between different batches of the medium already mentioned for gelatin probably also occurs with the agar. The differences shown in these tables between the counts obtained upon the different media are undoubtedly less than those that might be obtained with different batches of the same medium. The small number of tests made, therefore, can furnish only indications. To establish any actual difference between the various media would require a long series of tests, for which time was lacking in the present investigation. Lack of time has also made it impossible to test out any but the most significant points.

The first point tested was to determine the most satisfactory amount of asparaginate to use. It was found possible to vary this considerably without affecting the results. This is shown by the first four columns of Table IV. The usual formula containing 0.1

[23] This ordinarily requires just 10 c.c. of normal sodium hydroxide per litre.

TABLE IV.— TESTS TO DETERMINE THE EFFECT OF VARYING THE AMOUNT OF SODIUM ASPARAGINATE IN ASPARAGINATE AGAR.

Test No.	Date	Soil Type	Bacteria Per Gram Dry Soil, as Determined with —						
			Agar media as in Table I, column VI, containing —						Soil-extract gelatin.
			0.5 per ct. asparaginate.	0.2 per ct. asparaginate.	0.1 per ct. asparaginate.	0.05 per ct. asparaginate.	0.02 per ct. asparaginate.	No asparaginate.	
	1913.								
1	May 26	Dunkirk fine sandy loam		*15,000,000	13,500,000				12,000,000
2	June 2	Volusia silt loam		13,500,000	17,000,000				16,000,000
3	Sept. 6	Dunkirk gravelly loam	11,000,000		13,000,000				12,000,000
4	Sept. 15	Dunkirk loam	6,000,000		7,000,000				8,000,000
	1914.								
5	Jan. 16	Dunkirk silty clay loam			5,000,000	16,500,000			19,000,000
6	Jan. 16	Dunkirk silty clay loam			5,500,000	12,000,000			18,000,000
7	Jan. 16	Dunkirk silty clay loam			12,000,000	17,500,000			35,000,000
8	Jan. 19	Ontario fine sandy loam			15,000,000	20,000,000			27,000,000
9	Jan. 19	Ontario fine sandy loam			14,500,000	22,000,000			22,000,000
10	Jan. 24	Dunkirk silty clay loam			22,000,000	10,000,000			25,000,000
11	Jan. 24	Dunkirk silty clay loam			39,000,000	20,000,000			48,000,000
12	April 7	Dunkirk silty clay loam			24,000,000	23,000,000			30,000,000
13	April 7	Dunkirk silty clay loam			27,000,000	24,000,000			31,000,000
14	April 15	Dunkirk silty clay loam			13,000,000	16,000,000			23,000,000
15	April 24	Volusia silt loam			11,000,000	12,000,000			10,000,000
16	April 24	Volusia silt loam			15,000,000	18,000,000			12,000,000
17	April 24	Volusia silt loam			16,000,000	15,000,000			14,500,000
18	May 9	Dunkirk fine sand			10,500,000			9,500,000	10,000,000
19	May 14	Dunkirk silty clay loam			27,000,000			21,000,000	†20,000,000
20	May 14	Dunkirk silty clay loam			28,000,000			23,000,000	26,000,000
21	May 18	Ontario fine sandy loam			27,000,000			18,000,000	20,000,000
22	May 18	Ontario fine sandy loam			29,000,000	36,000,000			20,000,000
23	May 26	Volusia silt loam			9,000,000	9,000,000	6,000,000		†25,000,000

24	May 26	Volusia silt loam	......	......	**11,000,000**	**11,500,000**	......	23,000,000
25	May 28	Ontario fine sandy loam	......	20,000,000	......	19,000,000	......	21,000,000
26	May 28	Ontario fine sandy loam	......	18,000,000	......	17,000,000	......	21,000,000
27	May 29	Dunkirk silty clay loam	......	19,500,000	**13,000,000**	16,000,000	......	†20,000,000

* Counts upon the modified formulae that are higher than the corresponding counts upon the medium with 0.1 per ct. asparaginate are printed in bold-faced type.

† These counts are inexact because of rapid liquefaction.

per ct. asparaginate (see column three) was first compared with two modified formulæ containing 0.2 per ct. and 0.5 per ct. of asparaginate respectively (see columns one and two). As each of these formulæ was used in only two tests no definite conclusions can be drawn; but it is evident that there is no distinct advantage to be gained by using these larger quantities of asparaginate. A much longer series of tests was made of a medium containing only 0.05 per ct. asparaginate (see the fourth column). In ten of the sixteen cases the counts obtained with 0.05 per ct. asparaginate are higher than those obtained with 0.1 per ct.; while 0.1 per ct. of asparaginate has allowed better counts only in five cases. The slight superiority of the medium with the smaller amount of asparaginate may have been merely accidental; but it is quite plain that as good results can be obtained with 0.05 per ct. as with larger amounts. As the asparaginate is the most expensive constituent of the medium, perhaps it would be well in routine work to use 0.05 per ct. instead of 0.01 per ct. (as given in Table I). The results are less satisfactory, however, when the asparaginate has been lowered beneath this point. In the fifth column five counts on a medium containing only 0.02 per ct. sodium asparaginate are given, and in the sixth column four counts on a medium with the asparaginate omitted entirely. All these counts except one (test No. 24 with 0.02 per ct. asparaginate) were lower than the corresponding counts when 0.1 per ct. of asparaginate was used, but the differences were never very great. The greatest disadvantage of these two media cannot be shown by figures. Colonies developed very slowly upon them and remained small and undifferentiated in appearance. Although it is not absolutely necessary for a medium intended for quantitative work to show differences in appearance between the colonies of different kinds of bacteria, it is, nevertheless, a desirable feature if it can be obtained without sacrificing either of the other two more necessary qualifications. For this reason it is not advisable to use less than 0.05 per ct. of sodium asparaginate, even though it may be entirely omitted without greatly affecting the count.

A second series of tests, showing the effect of varying the dextrose content, is given in Table V. The six counts given in the first two columns indicate an inferiority when as much as 0.5 per ct. of dextrose is used, but the tests are too few in number to establish the fact. More work on this point is to be desired. It would be advantageous to use larger amounts of dextrose if it could be done without lowering the count, because this sugar is of considerable value in bringing out distinctions between the colonies of different kinds of bacteria. The counts given in the fourth column of this table would seem to show that reducing the dextrose content to 0.05 per ct. has lessened the number of colonies which develop; but

TABLE V.— TESTS TO DETERMINE THE EFFECT OF VARYING THE AMOUNT OF DEXTROSE IN ASPARAGINATE AGAR.

Test No.	Date.	Soil Type.	BACTERIA PER GRAM DRY SOIL, AS DETERMINED WITH—				
			ASPARAGINATE AGAR CONTAINING—				Soil-extract gelatin
			1 per ct. dextrose	0.5 per ct. dextrose	0.1 per ct. dextrose	0.05 per ct. dextrose	
	1913						
1	May 26	Dunkirk fine sandy loam		*10,000,000	13,500,000		12,000,000
2	June 2	Volusia silt loam		7,000,000	17,000,000		16,000,000
3	June 4	Dunkirk sandy loam		†3,000,000	7,000,000		7,500,000
4	June 24	Ontario loam			21,000,000	16,000,000	22,000,000
5	July 10	Dunkirk silty clay loam			17,500,000	12,500,000	22,000,000
6	Sept. 6	Dunkirk gravelly loam			13,000,000	11,000,000	12,000,000
	1914						
7	Oct. 27	Dunkirk silty clay loam	5,000,000		11,000,000		16,000,000
8	Oct. 30	Dunkirk silty clay loam	7,500,000		16,000,000		11,000,000
9	Oct. 30	Dunkirk silty clay loam	9,500,000		17,000,000		17,000,000

* The counts with these modified agar formulæ were always lower than the corresponding counts when 0.1 per ct. dextrose was employed; hence no bold-faced type is used in this table.
† The medium used in making this count contained 0.2 per ct. asparaginate.

some further tests are given in Table VI in which the use of 0.05 per ct. of dextrose has had no appreciable influence. This latter table contains thirteen comparative counts between the formula given in Table I (with 0.1 per ct. dextrose and 0.1 per ct. asparaginate) and a modified formula in which the asparaginate content is

TABLE VI.— TESTS TO DETERMINE THE EFFECT OF USING ONLY .05 PER CT. DEXTROSE IN ASPARAGINATE AGAR.

Test No.	Date.	SOIL TYPE.	BACTERIA PER GRAM DRY SOIL, AS DETERMINED WITH —		
			ASPARAGINATE AGAR CONTAINING —		Soil-extract gelatin
			0.1 per ct. dextrose	0.05 per ct. dextrose*	
	1913.				
1	Sept. 15	Dunkirk loam..............	7,000,000	†9,500,000	8,000,000
2	Sept. 24	Muck.....................	23,000,000	24,000,000	29,000,000
3	Oct. 4	Ontario fine sandy loam.....	4,500,000	7,000,000	7,000,000
4	Oct. 8	Volusia silt loam...........	4,500,000	9,500,000	9,300,000
5	Oct. 8	Volusia silt loam...........	4,000,000	7,500,000	9,800,000
6	Oct. 22	Genesee loam..............	19,000,000	22,500,000	27,000,000
7	Oct. 27	Muck.....................	74,000,000	70,000,000	118,000,000
8	Nov. 18	Dunkirk fine sand.........	6,500,000	7,800,000	8,000,000
9	Nov. 18	Dunkirk fine sand.........	6,000,000	7,500,000	9,000,000
10	Nov. 25	Dunkirk silty clay loam.....	9,500,000	10,500,000	16,000,000
11	Dec. 5	Ontario fine sandy loam.....	18,000,000	‡17,000,000	19,000,000
12	Dec. 5	Ontario fine sandy loam.....	18,000,000	17,000,000	24,000,000
13	Dec. 15	Dunkirk silty clay loam.....	12,500,000	10,500,000	12,000,000

* This medium contained 0.2 per ct. asparaginate.

† Counts on the asparaginate agar with 0.05 per ct. dextrose that are higher than the corresponding counts upon the medium with 0.1 per ct. dextrose are printed in bold-faced type.

‡In this case there was such irregularity between the number of colonies upon the parallel plates that a satisfactory average could not be taken.

doubled but only half the usual amount of dextrose is used. The count has proved higher on the modified formula than on the ordinary asparaginate agar in all but four cases, and equal to it in one of those four; but the counts on the two media are always so nearly the same that no weight can be attached to the differences between them. It is plain that, if reducing the amount of dextrose has had any influence upon the count, that influence has been neutralized by increasing the amount of asparaginate, a result which is not to be expected in view of the data given in Table IV, showing that variations in the amount of asparaginate have no appreciable effect upon the number of colonies that develop. For quantitative purposes, 0.05 per ct. dextrose seems to be as good as 0.1 per ct. Possibly the dextrose could be entirely omitted without causing a lower count. In fact a single test in which a formula was used differing from that of Table I only in the absence of dextrose, resulted in exactly the same count as obtained with the use of 0.1 per ct. dextrose.[24] No further tests were made with this formula as the colonies developing on it were all very small and alike in appearance.

A further series of tests was made to see if the formula of this medium could be simplified. In Table VII the counts obtained by the use of three agar media of more simple composition are compared with those made on the ordinary formula. One of the simpler media is a mixture of agar, tap-water, 0.1 per ct. of sodium asparaginate and 0.1 per ct. of dextrose; the second the same with the dextrose omitted; and the third a mixture of agar and tap-water alone. This comparison was made in the hope that tap-water might supply all the necessary mineral salts. Only two tests were made of tap-water and agar alone, because, although this medium allowed a fairly high count, the colonies were all small and of the same appearance. When sodium asparaginate was added to this tap-water agar, however, the results were fairly satisfactory, and with the further addition of dextrose more satisfactory still, but the colonies were not even then as large as when the formula given in Table I was used, and the count was usually lower. This attempt at simplification cannot be considered a success.

It has already been mentioned that a most important point in the composition of the asparaginate agar is its reaction. There is very little data available to prove this point by direct comparison, although it has been well established in the course of the present work. The fact was learned largely by noticing that whenever the medium was made up by accident with a reaction as high as 1.5 per ct. normal acid, the counts obtained were always much lower than expected. This was so very evident that a few additional direct tests were considered enough to settle the matter. They are given in Table VIII. The four tests all agree in showing that the count is about twice as high when the reaction is 0.8 per ct. as when it is 1.5 per ct. Additional weight is given to these figures by the fact that tests Nos. 1 and 2 were made with different batches of media from those used in Nos. 3 and 4, the media used in the last two tests having a slightly different formula from usual (0.05 per ct. dextrose and 0.2 per ct. asparaginate). No media were tested out with a reaction more alkaline than 0.8 per ct. acid, because of the danger of losing the ammonia of ammonium phosphate unless the medium was

---

[24] This test is not included in any of the tables.

TABLE VII.— TESTS OF SIMPLIFIED FORMULAE FOR ASPARAGINATE AGAR.

Test No.	Date.	Soil Type.	BACTERIA PER GRAM DRY SOIL, AS DETERMINED WITH —				
			Asparaginate agar, as given in Table I.	SIMPLIFIED FORMULAE, CONTAINING —			Soil-extract gelatin.
				Agar, tap-water, sodium asparaginate and dextrose.	Agar, tap-water, and sodium asparaginate.	Agar and tap-water.	
	1914.						
1	May 9	Dunkirk fine sand	10,500,000			*11,000,000	10,000,000
2	May 9	Dunkirk fine sand				10,000,000	11,000,000
3	May 19	Dunkirk sandy loam	6,000,000		5,000,000		4,000,000
4	May 19	Dunkirk sandy loam	7,500,000		7,000,000		‡9,000,000
5	May 21	Muck	160,000,000		**210,000,000**		180,000,000
6	May 21	Muck	92,000,000		**100,000,000**		‡100,000,000
7	Aug. 27	Dunkirk silty clay loam	20,000,000	16,000,000	16,000,000		†17,000,000
8	Aug. 27	Dunkirk silty clay loam	16,000,000	9,700,000	9,200,000		†10,000,000
9	Aug. 28	Ontario loam	33,000,000	23,000,000	22,000,000		†23,000,000
10	Aug. 28	Ontario loam	42,000,000	22,000,000	18,500,000		†32,000,000
11	Aug. 31	Dunkirk silty clay loam	25,000,000	21,000,000	19,000,000		†20,000,000
12	Aug. 31	Dunkirk silty clay loam	20,000,000	16,000,000	15,000,000		†19,000,000
13	Sept. 1	Dunkirk silty clay loam	36,000,000	28,000,000	33,000,000		†35,000,000
14	Sept. 1	Dunkirk silty clay loam	22,000,000	16,000,000	15,000,000		24,000,000
15	Sept. 2	Dunkirk silty clay loam	15,000,000	12,000,000	11,500,000		14,000,000
16	Sept. 2	Dunkirk silty clay loam	13,000,000	9,200,000	9,000,000		12,000,000
17	Sept. 5	Dunkirk silty clay loam	30,000,000	24,000,000	24,000,000		35,000,000
18	Sept. 5	Dunkirk silty clay loam	21,000,000	17,000,000	17,000,000		24,000,000

* Counts on the simplified formulæ that are higher than the corresponding counts upon the full formula for asparaginate agar are printed in bold-faced type.

† Counts made with tap-water gelatin, instead of soil-extract gelatin.

‡ These counts are inexact because of rapid liquefaction.

distinctly acid in reaction. These tests bear out the generally admitted fact that media for soil work should have a reaction between 0.5 per ct. and 1.0 per ct. normal acid to phenolphthalein.

TABLE VIII.— TESTS TO DETERMINE THE EFFECT OF VARYING THE REACTION OF ASPARAGINATE AGAR.

Test No.	Date.	SOIL TYPE.	BACTERIA PER GRAM DRY SOIL, AS DETERMINED ON ASPARAGINATE AGAR WITH A REACTION OF	
			0.8 per ct. acid.	1.5 per ct. acid.
	1913.			
1	Dec. 5	Ontario fine sandy loam........	18,000,000	8,500,000
2	Dec. 5	Ontario fine sandy loam........	18,000,000	7,500,000
3*	Dec. 5	Ontario fine sandy loam........	17,000,000	7,000,000
4*	Dec. 5	Ontario fine sandy loam........	17,000,000	9,000,000

* Tests Nos. 3 and 4 were made with different batches of media from the first two tests. The media used in Nos. 3 and 4 contained 0.05 per ct. dextrose and 0.2 per ct. asparaginate, instead of the usual amounts.

TESTS COMPARING THE VARIOUS MEDIA.

A series of tests was made comparing the soil extract gelatin and the asparaginate agar with the other solid media that have been recommended for soil bacteria. Table IX is a comparison between the counts obtained upon the asparaginate agar and parallel counts upon the soil-extract gelatin. In Tables X to XIII the counts upon these two media are compared with those upon the media recommended by Fischer, by Lipman and Brown, by Temple and by Brown.

In fifty-nine of the ninety-six comparative tests given in Table IX, the soil-extract gelatin gave higher counts than the asparaginate agar. In thirty-four cases the counts upon the agar were higher, and in three cases both media gave the same count. These tests show that the gelatin medium is rather better than the agar medium if we judge by the number of soil bacteria that grow upon it. In the matter of distinctions in appearance between colonies of different bacteria, it has already been stated that the gelatin is the more satisfactory; but the requirement of definite chemical composition is more nearly met by the agar. From these facts it may be concluded that the gelatin is best for qualitative work, the agar best for quantitative work. One other disadvantage of the gelatin, its rapid liquefaction by certain organisms, constitutes a further

TABLE IX.— TESTS COMPARING THE SOIL-EXTRACT GELATIN WITH THE ASPARAGINATE AGAR.

Test No.	Date	Soil Type	Bacteria Per Gram Dry Soil, as Determined with —	
			Soil-extract gelatin	Asparaginate agar
	1913.			
1	April 4	Dunkirk silty clay loam	*27,000,000	27,000,000
2	April 5	Volusia silt loam	19,000,000	26,000,000
3	April 14	Dunkirk silty clay loam	28,000,000	25,000,000
4	April 14	Dunkirk silty clay loam	48,000,000	47,000,000
5	April 21	Dunkirk silty clay loam	33,000,000	29,000,000
6	April 24	Ontario fine sandy loam	23,000,000	24,500,000
7	May 13	Honeoye stony loam	28,000,000	29,000,000
8	May 26	Dunkirk fine sandy loam	12,000,000	13,500,000
9	June 4	Volusia silt loam	16,000,000	17,000,000
10	June 4	Dunkirk sandy loam	7,500,000	7,000,000
11	June 26	Volusia silt loam	15,000,000	17,500,000
12	July 10	Vclusia silt loam	22,000,000	16,600,000
13	July 14	Dunkirk silty clay loam	15,000,000	26,500,000
14		Muck	35,000,000	13,000,000
15	Sept. 15	Dunkirk gravelly loam	12,000,000	7,000,000
16	Sept. 24	Dunkirk loam	8,000,000	23,000,000
17		Muck	29,000,000	10,000,000
18	Oct. 8	Ontario loam	7,000,000	4,500,000
19	Oct. 8	Volusia silt loam	8,700,000	5,000,000
20	Oct. 8	Volusia silt loam	9,300,000	4,500,000
21	Oct. 8	Volusia silt loam	9,000,000	5,000,000
22	Oct. 22	Genesee loam	9,800,000	4,000,000
23	Oct. 22	Genesee loam	30,000,000	18,500,000
24	Oct. 27	Muck	23,500,000	19,000,000
25		Muck	118,000,000	74,000,000
26	Nov. 18	Dunkirk fine sand	9,000,000	6,500,000
27	Nov. 18	Dunkirk fine sand	8,000,000	6,000,000
28	Nov. 20	Ontario fine sandy loam	27,500,000	†20,000,000
29	Nov. 20	Ontario fine sandy loam	21,000,000	‡18,000,000
30	Nov. 25	Dunkirk silty clay	16,000,000	10,000,000
31	Dec. 1	Dunkirk fine sand	8,800,000	8,300,000
32	Dec. 1	Dunkirk fine sand	7,500,000	7,540,000
33	Dec. 5	Ontario fine sandy loam	19,000,000	18,000,000
34	Dec. 5	Ontario fine sandy loam	24,000,000	18,000,000
35	Dec. 11	Dunkirk silty clay loam	16,000,000	16,000,000
	Dec. 11	Dunkirk silty clay loam	22,000,000	‡18,000,000
49	Feb. 20	Dunkirk silty clay loam	58,000,000	65,500,000
50	Feb. 20	Dunkirk silty clay loam	48,000,000	75,000,000
51	Feb. 26	Dunkirk silty clay loam	38,000,000	44,000,000
52	Feb. 26	Dunkirk silty clay loam	59,000,000	52,000,000
53	Feb. 27	Dunkirk fine sand	57,000,000	90,000,000
54	Feb. 28	Dunkirk fine sand	95,000,000	97,000,000
55	Feb. 28	Dunkirk silty clay loam	42,000,000	54,000,000
56	Feb. 28	Dunkirk silty clay loam	71,000,000	65,000,000
57	Mar. 14	Dunkirk fine sand	8,000,000	12,500,000
58	Mar. 14	Dunkirk fine sand	26,000,000	25,500,000
59	Mar. 17	Volusia silt loam	14,000,000	15,500,000
60	April 7	Dunkirk silty clay loam	30,000,000	24,000,000
61	April 7	Dunkirk silty clay loam	31,000,000	27,000,000
62	April 15	Dunkirk silty clay loam	21,500,000	13,000,000
63	April 15	Dunkirk silty clay loam	23,000,000	13,000,000
64	April 15	Dunkirk silty clay loam	32,000,000	19,000,000
65	April 24	Volusia silt loam	10,000,000	11,000,000
66	April 24	Volusia silt loam	12,000,000	15,000,000
67	April 24	Volusia silt loam	14,500,000	16,000,000
68	May 14	Dunkirk fine sand	10,000,000	10,500,000
69	May 14	Dunkirk silty clay loam	†20,000,000	27,000,000
70	May 18	Dunkirk silty clay loam	26,000,000	28,000,000
71	May 18	Ontario fine sandy loam	20,000,000	27,000,000
72	May 19	Ontario fine sandy loam	†25,000,000	29,000,000
73	May 19	Dunkirk sandy loam	4,000,000	6,000,000
74	May 21	Dunkirk sandy loam	†9,000,000	7,500,000
75		Muck	180,000,000	160,000,000
76		Muck	†100,000,000	92,000,000
77	May 28	Ontario fine sandy loam	23,000,000	20,000,000
78	May 28	Ontario fine sandy loam	21,000,000	18,000,000
79	May 29	Dunkirk silty clay loam	†20,000,000	19,500,000
80	Aug. 5	Volusia silt loam	9,500,000	8,000,000
81	Aug. 5	Volusia silt loam	12,500,000	7,500,000
82	Aug. 6	Volusia silt loam	8,300,000	11,500,000
83	Aug. 6	Volusia silt loam	10,000,000	12,000,000

No.	Date	Soil		
36	Dec. 15	Dunkirk silty clay loam....	12,000,000	**12,500,000**
37	Dec. 15	Dunkirk silty clay loam....	**14,000,000**	10,500,000
	1914.			
38	Jan. 16	Dunkirk silty clay loam....	**19,000,000**	5,000,000
39	Jan. 16	Dunkirk silty clay loam....	**18,000,000**	5,500,000
40	Jan. 16	Dunkirk silty clay loam....	**35,000,000**	12,000,000
41	Jan. 19	Ontario fine sandy loam....	**27,000,000**	15,000,000
42	Jan. 19	Ontario fine sandy loam....	**22,000,000**	14,500,000
43	Jan. 24	Dunkirk silty clay	**25,000,000**	22,000,000
44	Jan. 24	Dunkirk silty clay loam....	**48,000,000**	39,000,000
45	Jan. 28	Dunkirk silty clay loam....	†60,000,000	**70,000,000**
46	Jan. 28	Dunkirk silty clay loam....	**56,000,000**	36,000,000
47	Jan. 30	Dunkirk silty clay loam....	**33,000,000**	22,500,000
48	Jan. 30	Dunkirk silty clay loam....	**34,000,000**	15,000,000

No.	Date	Soil		
84	Aug. 6	Volusia silt loam......	**9,500,000**	6,700,000
85	Aug. 7	Dunkirk silty clay loam..	§9,000,000	**14,500,000**
86	Aug. 7	Dunkirk silty clay loam.	10,000,000	**11,000,000**
87	Aug. 8	Ontario loam....	**21,000,000**	9,000,000
88	Aug. 8	Ontario loam....	**25,000,000**	16,000,000
89	Aug. 10	Volusia silt loam....	5,000,000	**5,500,000**
90	Aug. 10	Volusia silt loam....	**8,500,000**	7,000,000
91	Aug. 11	Volusia silt loam....	**10,500,000**	6,000,000
92	Sept. 1	Dunkirk silty clay loam..	**24,000,000**	22,000,000
93	Sept. 2	Dunkirk silty clay loam..	14,000,000	**15,000,000**
94	Sept. 2	Dunkirk silty clay loam.	12,000,000	**13,000,000**
95	Sept. 5	Dunkirk silty clay loam.	**35,000,000**	30,000,000
96	Sept. 5	Dunkirk silty clay loam..	**24,000,000**	21,000,000

* The higher count in each test is printed in bold-faced type.
† These counts are inexact because of rapid liquefaction.
‡ The medium used in making these counts contained 0.05 per ct. dextrose and 0.2 per ct. sodium asparaginate.
§ The medium used in making this count was tap-water gelatin without dextrose.

objection to its use in quantitative work. Although the liquefaction is slower than on beef-extract-peptone gelatin, still at times it proceeds so rapidly as to prevent any count. The most efficient method found to inhibit the growth of the liquefiers without stopping the growth of other bacteria is to use an incubation temperature that does not exceed 18° C. It seems possible, indeed, that the rapid liquefaction which has so often led soil bacteriologists to regard gelatin with disfavor may have resulted from their use of a temperature of 20–21° C. for incubation. With the use of a sufficiently low temperature there has seldom been any great trouble in keeping the gelatin plates seven days before counting. Low temperatures are advisable whether the medium is to be used for qualitative or quantitative purposes, although more necessary in the latter case than in the former.[25] The asparaginate agar, on the other hand, can be used even when low temperatures are unavailable.

TABLE X.— TESTS OF FISCHER'S CULTURE MEDIUM.

Test No.	Date.	SOIL TYPE.	BACTERIA PER GRAM DRY SOIL, AS DETERMINED WITH —		
			Asparaginate agar.	Fischer's agar.	Soil-extract gelatin.
	1913.				
1	Apr. 21	Dunkirk silty clay loam..	29,000,000	42,000,000	33,000,000
2	Oct. 4	Ontario fine sandy loam..	4,500,000	11,500,000	7,000,000
3	Oct. 8	Volusia silt loam........	5,000,000	16,000,000	8,700,000
4	Nov. 24	Dunkirk silty clay loam..	†14,000,000	17,000,000	21,000,000
5	Nov. 25	Dunkirk silty clay loam..	9,500,000	13,000,000	16,000,000
6	Dec. 26	Dunkirk silty clay loam..	*26,000,000	25,000,000	Liquefied
7	Dec. 26	Dunkirk silty clay loam..	37,000,000	32,000,000	Liquefied

* Counts upon the asparaginate agar and upon soil-extract gelatin that are higher than the corresponding counts upon Fischer's agar are printed in bold-faced type.

† The medium used in making this count contained 0.2 per ct. asparaginate and only 0.05 per ct. dextrose.

A comparison between the counts obtained on the gelatin and on the other soil media may be obtained from the figures given in Tables X to XIII. Fischer's agar, in the series of tests listed in Table X, gave a higher count than the gelatin three times out of five, in those listed in Table XIII only four times out of twelve. Lipman and Brown's agar gave a higher count than the gelatin in

[25] Liquefaction may also be checked by using 20 per ct. instead of 12 per ct. gelatin. This does not seem to lower the number of colonies.

only two of the twenty-two tests listed in Tables XI and XIII.
Brown's agar gave as high a count as the gelatin only once in the
nine tests given in Table XII, and four times in the twelve tests
of Table XIII. Temple's agar gave a higher count than the gelatin
in just three of the twelve tests listed in Table XIII. These counts
show that these four agar media, like the asparaginate agar, permit
the growth of fewer soil bacteria than does the gelatin. None of
them allow as good distinction in appearance between the different

TABLE XI.— TESTS OF LIPMAN AND BROWN'S CULTURE MEDIUM.

Test No.	Date.	SOIL TYPE.	BACTERIA PER GRAM DRY SOIL, AS DETERMINED WITH —		
			Asparaginate agar.	Lipman and Brown's agar.	Soil-extract gelatin.
	1913.				
1	Apr. 21	Dunkirk silty clay loam..	*29,000,000	20,000,000	33,000,000
2	Jan. 16	Dunkirk silty clay loam..	†16,500,000	7,500,000	19,000,000
3	Jan. 16	Dunkirk silty clay loam..	†12,000,000	6,500,000	18,000,000
4	Jan. 16	Dunkirk silty clay loam..	†17,00,000	8,800,000	35,000,000
	1914.				
5	Jan. 19	Ontario fine sandy loam..	15,000,000	15,800,000	27,000,000
6	Jan. 19	Ontario fine sandy loam..	14,500,000	15,000,000	22,000,000
7	Jan. 24	Dunkirk silty clay loam..	22,000,000	13,000,000	25,000,000
8	Jan. 24	Dunkirk silty clay loam..	39,000,000	16,500,000	48,000,000
9	Jan. 28	Dunkirk silty clay loam..	70,000,000	35,000,000	‡60,000,000
10	Jan. 28	Dunkirk silty clay loam..	36,000,000	18,000,000	56,000,000

* Counts upon asparaginate agar and upon soil-extract gelatin that are higher than
the corresponding counts upon Lipman and Brown's agar are printed in bold-faced
type.
† The medium used in making these counts contained only 0.05 per ct. asparaginate.
‡ This count is inexact because of rapid liquefaction.

colonies as does gelatin. With the exception of Lipman and Brown's
agar, none of them have any advantage over the gelatin in the
matter of definite chemical composition.

The same tables show how the counts obtained upon the asparaginate agar compare with those obtained upon the other four agar
media. Fischer's agar gave a higher count than the asparaginate
agar in five out of the seven tests listed in Table X, but in only four
of the twelve tests included in Table XIII. Lipman and Brown's
agar gave higher counts than the asparaginate agar in only four of
the twenty-two tests included in Tables XI and XIII, and then
always by a very narrow margin; while in several of the tests in which
it has given a lower count than the asparaginate agar (as in the last

four tests of Table XI) the difference has been very pronounced. Brown's agar has given slightly higher counts than the asparaginate agar in three of the nine tests of Table XII, but in the other six tests has given much lower counts than the asparaginate agar; and in Table XIII has given higher counts than on the asparaginate agar in five of the twelve tests. Temple's agar has given higher counts than the asparaginate agar in six of the twelve tests listed in Table XIII. These counts show that the asparaginate agar is adapted to the growth of at least as large a number of soil bacteria as any of the other agar media; in this respect it is superior to them rather than inferior, and is unquestionably superior to Lipman and Brown's

TABLE XII.— TESTS OF BROWN'S CULTURE MEDIUM.

Test No.	Date.	SOIL TYPE.	BACTERIA PER GRAM DRY SOIL, AS DETERMINED WITH —		
			Asparaginate agar.	Brown's agar.	Soil-extract gelatin.
	1913.				
1	Sept. 24	Muck..................	*24,000,000	†13,000,000	29,000,000
2	Oct. 4	Ontario fine sandy loam..	4,500,000	7,000,000	7,000,000
3	Oct. 8	Volusia silt loam........	5,000,000	†8,000,000	8,700,000
4	Oct. 8	Volusia silt loam........	5,000,000	†4,500,000	9,000,000
5	Oct. 8	Volusia silt loam........	4,000,000	6,000,000	9,800,000
6	Oct. 22	Genesee loam...........	19,000,000	15,500,000	27,000,000
7	Nov. 20	Ontario fine sandy loam..	‡20,000,000	14,000,000	27,000,000
8	Nov. 20	Ontario fine sandy loam..	‡18,000,000	10,000,000	21,000,000
9	Dec. 15	Dunkirk silty clay loam..	10,500,000	5,000,000	14,000,000

* Counts upon the asparaginate agar and upon soil-extract gelatin that are higher than the corresponding counts upon Brown's agar are printed in bold-faced type.

† In these cases there was such irregularity between the counts from the parallel plates that a satisfactory average could not be taken.

‡ The medium used in making these counts contained 0.2 per ct. asparaginate and only 0.05 per ct. dextrose.

agar. It has been found to allow much greater distinction in appearance between different kinds of colonies than Fischer's agar, slightly greater than Temple's, while it is certainly not inferior in this matter to Lipman and Brown's or to Brown's formula. In respect to definite chemical composition, as already stated, it is superior to all four.

Table XIII is of particular interest because all six of these media were included in this series of comparative tests. In these twelve tests there was very little variation between the counts obtained upon the different media. They do not even show the usual superi-

TABLE XIII.— Tests Comparing the Six Culture Media Described in Table I.

Test No.	Date.	Soil Type.	Bacteria Per Gram Dry Soil, as Determined with —					
			Soil-extract gelatin.	Asparaginate agar.	Lipman and Brown's agar.	Fischer's agar.	Brown's agar.	Temple's agar.
	1914.							
1	Aug. 5	...lia silt loam	9,500,000	8,000,000	5,500,000	6,700,000	8,000,000	*8,800,000
2	Aug. 5	Volusia silt loam	12,500,000	7,500,000	5,300,000	5,000,000	7,000,000	7,200,000
3	Aug. 6	Volusia silt loam	8,300,000	†11,500,000	11,500,000	10,500,000	11,000,000	9,300,000
4	Aug. 6	...Uia silt loam	10,000,000	12,000,000	8,800,000	12,000,000	11,000,000	12,000,000
5	Aug. 6	Volusia silt loam	9,500,000	6,700,000	6,000,000	*8,500,000	*8,300,000	*6,800,000
6	Aug. 7	...k silty clay loam	9,000,000	14,500,000	14,000,000	11,800,000	*15,000,000	*15,000,000
7	Aug. 7	Dunkirk silty clay loam	10,000,000	11,000,000	9,600,000	8,500,000	8,800,00	10,000,000
8	Aug. 8	Ontario loam	21,000,000	9,000,000	*13,000,000	*15,000,000	*14,000,000	*16,000,000
9	Aug. 8	Ontario loam	25,000,000	16,000,000	16,000,000	*21,000,000	*17,000,00	*20,000,000
10	Aug. 10	Volusia silt loam	5,000,000	5,500,000	5,500,000	5,400,000	5,50,000	5,000,000
11	Aug. 10	...lia silt loam	8,500,000	7,000,000	6,600,000	5,800,000	5,20,000	6,700,000
12	Aug. 11	...lia silt loam	10,500,000	6,000,000	*9,800,000	*9,000,000	*6,300,00	*7,000,000

* Counts on the last four media mentioned that are higher than the corresponding counts on asparaginate agar.
† Agar counts that are higher than the corresponding count on soil-extract gelatin are printed in bold-faced type.

ority of gelatin so far as count is concerned, as in only two tests (Nos. 8 and 9) was the gelatin count appreciably different from the agar counts. Sometimes one medium has given the highest count, sometimes another. None of them has given consistently lower counts than the asparaginate agar, with the possible exception of Lipman and Brown's medium, which has given a higher count only twice, and has equalled the asparaginate agar in count only three other times. These results, considered together with those listed in Table XI, give a particularly unfavorable showing to Lipman and Brown's agar, a fact which is important when it is remembered that this medium is the only one except the asparaginate agar that does not contain an appreciable amount of any substance of indefinite chemical composition. The differences between the various counts in Table XIII, however, are all too small to be of significance; and it must be concluded that under favorable conditions any one of these media (with the possible exception of Lipman and Brown's) is adapted to the growth of as many soil bacteria as any of the others. In choosing between them, the decision must be based upon other matters rather than upon the number of colonies they allow to develop.

Important considerations to be taken into account are these: A great drawback of Fischer's agar is that the colonies are all mere pin-points and cannot be distinguished from one another. A serious disadvantage of Lipman and Brown's medium and of Brown's modification of it is that molds and overgrowths are often so abundant upon them as to interfere with the counting and prevent the isolation of pure cultures from the colonies. A further objection to Brown's agar arises from the difficulty of obtaining an even distribution of the albumin, which must be added after the medium has cooled enough not to cause coagulation but before it is cold enough to prevent tubing. Temple's agar proves especially attractive to *Bacillus mycoides*, which was so abundant and vigorous in some of the soils studied as to overgrow the plates and to render counting difficult. Considering these points in addition to the advantages of the two new media that have already been discussed — the superiority of gelatin in the matter of allowing distinctions in appearance between different colonies, and of the asparaginate agar in the matter of definite composition — it must be concluded that the gelatin is the best medium for qualitative purposes, the asparaginate agar for quantitative work.

CONCLUSIONS.

Determinations of the number of bacteria present in soil are generally made by counting the colonies that develop on gelatin or agar media. Results depend largely upon the composition of the medium. Three important characteristics are to be looked for in a medium for this purpose; it should allow the greatest possible number of soil bacteria to develop upon it, in order that the counts obtained may be as nearly correct as possible; it should allow the different kinds of bacteria to produce colonies as distinct as possible in appearance, in order to facilitate classification; and it should contain, as far as possible, no materials of unknown chemical formula, in order that different batches of the medium may be of the same chemical composition. In the past, these last two requirements have been almost overlooked.

In the course of the present investigation it has not been found possible to obtain a medium fulfilling all three of these requirements; but two new media have been tested out that are worth recommending. Both of these media fulfill the first requirement as well as any previously proposed medium. One of them, a soil-extract gelatin, fulfills the second requirement better than any of the other media proposed for soil work, and is therefore recommended for qualitative work. The other, an agar containing sodium asparaginate, fulfills the third requirement (except in so far as the agar itself is of indefinite composition) and is therefore recommended for quantitative work.

The soil-extract gelatin consists of gelatin, soil-extract and dextrose alone. This medium not only permits more ready classification of the colonies than any other medium tested, but also allows a larger number of colonies to develop than appear on the media ordinarily recommended for soil work. The soil-extract is not absolutely necessary, as practically as good results may be obtained when it is replaced with tap-water, and only slightly inferior results when distilled water is used in its stead. This gelatin medium is extremely satisfactory for qualitative work; and might also be recommended for quantitative work except for the indefinite composition of the gelatin itself.

The asparaginate agar contains no organic matter except the agar, dextrose and sodium asparaginate. Besides these materials, it contains several mineral salts. The formula given in this bulletin is not to be considered the best combination possible, although various proportions of the different chemicals have been tried without better success. Nearly as good quantitative results may be obtained by omitting the mineral salts and using tap-water instead of distilled water; but the colonies then are so small that this simplification is not to be recommended. The colonies developing on asparaginate agar are not so readily classified as those on gelatin, and the count obtained upon it is often lower, but its definite chemical composition makes it seem worth recommending for general use in quantitative work in soil bacteriology.

The media that have been compared with these are: Fischer's soil-extract agar; Temple's peptone agar; Lipman and Brown's "synthetic agar" containing peptone, and Brown's modification of the latter in which the peptone is replaced by albumin. For qualitative work none of these media is as good as gelatin, but all of them except Fischer's allow some differences in appearance between the colonies of different kinds of bacteria. For quantitative work all four are about equally satisfactory; none of them give higher counts than the asparaginate agar.

Three points are brought out plainly by this investigation: (1) Gelatin media are not only better than agar media for qualitative work, but allow as many if not more of the soil bacteria to produce colonies. (2) A satisfactory agar medium can be prepared containing nothing of indefinite chemical composition except the agar itself. (3) This agar medium and those agar media especially recommended by Fischer, by Temple, by Lipman, and by Brown all give quantitative results so nearly alike that the counts obtained on any one of them may be compared with those obtained on any other, provided the same technique of incubation be used.

ACKNOWLEDGMENTS.

This work was begun while Dr. H. A. Harding was bacteriologist at this Station, and was completed under the direction of Dr. R. S. Breed, the present bacteriologist. Acknowledgments are due to them for the assistance they have given.

TECHNICAL BULLETIN No. 39. DECEMBER, 1914.

New York Agricultural Experiment Station.

GENEVA, N. Y.

CONDITION OF CASEIN AND SALTS IN MILK.

LUCIUS L. VAN SLYKE AND ALFRED W. BOSWORTH.

PUBLISHED BY THE DEPARTMENT OF AGRICULTURE.

CONDITION OF CASEIN AND SALTS IN MILK.

LUCIUS L. VAN SLYKE AND ALFRED W. BOSWORTH.

SUMMARY.

1. Milk contains two general classes of compounds, those in true solution and those in suspension, or insoluble. These two portions can be separated for study by filtering the milk through a porous earthenware filter like the Pasteur-Chamberland filtering tube.

2. Serum prepared from fresh milk is yellow with a faint greenish tinge and slight opalescence. The following constituents of milk are wholly in solution in the milk-serum: Sugar, citric acid, potassium, sodium and chlorine. The following are partly in solution and partly in suspension: Albumin, inorganic phosphates, calcium, magnesium. Albumin in fresh milk appears to be adsorbed to a considerable extent by casein and therefore only a part of it appears in the serum. In serum from sour milk and milk to which formaldehyde has been added, nearly all of the albumin appears in the serum.

3. The insoluble portion of milk separated by filtration through the Pasteur-Chamberland filtering tube is grayish to greenish white in color, of a glistening, slime-like appearance and gelatinous consistency. When shaken with water it goes readily into suspension, forming a mixture having the opaque, white appearance of milk. Such a suspension is neutral to phenolphthalein. When purified, the insoluble portion consists of neutral calcium caseinate (casein Ca_4) and neutral di-calcium phosphate ($CaHPO_4$). The casein and di-calcium phosphate are not in combination, as shown by a study of 16 samples of milk from 13 individual cows, and also by a study of the deposit or " separator slime " formed by whirling milk in a cream separator. By treating fresh milk with formaldehyde and whirling in a centrifugal machine under specified conditions, it is possible to effect a nearly complete separation of phosphates from casein.

4. Both fresh milk and the serum from fresh milk show a slight acid reaction to phenolphthalein but are strongly alkaline to methyl orange, indicating that acidity is due, in part at least, to acid phosphates. In 8 samples of fresh milk, the acidity of the milk and of the milk-serum was determined after treatment with neutral potassium oxalate. The results show that that acidity of the whole milk is the same as that of the serum and that, therefore, the con-

stituents of the serum are responsible for the acidity of milk. There is every reason to believe that the phosphates of the serum cause the observed acidity.

5. The data presented, with results of other work, furnish a basis for suggesting an arrangement of the individual compounds contained in milk, especially including the salts.

INTRODUCTION.

The chemistry of milk has been studied by many investigators. Numerous facts have been accumulated relating to the amounts and properties of the more prominent constituents of milk, including various conditions affecting the composition; but much less attention has been given to thorough study of individual constituents, owing largely to the difficulties involved in making such investigations.

From the beginning of its existence, this Station has given much attention to study of different phases of the composition of milk. In connection with the study of the relation of the constituents of milk to cheese-making, to fermented beverages made from milk, and to the uses of milk in human nutrition, numerous chemical questions have constantly arisen and continue to come up, to which satisfactory answers can not be given, owing to our lack of knowledge of the chemistry of some of the milk constituents. Until our knowledge in this field becomes more complete, we cannot understand fully, for example, the fundamental chemical facts involved in the process of cheese-making and cheese-ripening, the chemical changes taking place in its constituents when milk sours or when it is made into fermented beverages such as kumyss, imitation butter-milks, matzoon, zoolak, bulgarzoon, etc.

We have in hand investigations relating to several of the fundamental questions referred to. In the present bulletin, we shall present the results of our work bearing on the following points:

(1) Properties and composition of milk serum or constituents in solution.

(2) Properties and composition of portion of constituents not in solution.

(3) Acidity of milk and milk-serum.

(4) The salts of milk.

METHOD OF PREPARING MILK-SERUM.

Before taking up the detailed results relating to these lines of investigation, we will give a description of the method used in preparing milk-serum from milk.

That portion of the milk consisting of water and the compounds in solution is known as the milk-serum. In studying the individual constituents of milk, it is necessary to separate the serum. Various methods have been used to separate milk-serum from the other constituents of milk, but the one best adapted for investigational purposes depends upon the fact that when milk is brought into contact with a porous earthenware filter, the water passes through, carrying with it the compounds in true solution, while the compounds insoluble in water or in suspension remain on the surface of the filter. In one form or another, this fact has been utilized in studying milk by Lehman, Duclaux, Eugling, Söldner and others. The form of earthenware filter used by us is much superior to any employed by these investigators. We have made use of the special form of apparatus designed by Briggs[1] for the purpose of obtaining water-extracts from soils. Briefly stated, the process consists in putting the milk to be examined into a tubular chamber surrounding a Pasteur-Chamberland filtering tube; pressure, amounting to 40 to 45 pounds per square inch, is applied by means of a pump which forces air into the chamber containing the milk and causes the soluble portion of the milk to pass through the walls of the filter from the outside to the inside of the filtering tube, from which it runs out and is caught in a flask standing underneath. The insoluble residue accumulates on the outside surface of the filter tube from which it can easily be removed by light scraping. .

It has been found by Rupp[2] that the filter appears to have the power of absorbing some of the soluble constituents of the serum until a volume of 50 to 75 c.c. has passed through, after which the filtered serum is constant in composition. In our work, therefore, the first portion of serum filtered is not used.

Before being placed in the apparatus for filtration, the milk is treated with some antiseptic to prevent souring during the process of filtration.

The composition of the solid portion of milk removed by the filtering tube is ascertained by difference; from the figures obtained by an analysis of the original milk we subtract the results of analysis given by the serum.

PROPERTIES AND COMPOSITION OF MILK-SERUM.

Serum prepared from fresh milk by the method described above has a characteristic appearance, being of a yellow color with a faint greenish tinge and slight opalescence.

The serum from fresh milk gives a slight acid reaction to phenolphthalein and a strongly alkaline reaction to methyl orange. We will later give the results of a special study made of the cause of acidity in milk-serum.

In the table below we give the results of the examination of two samples of fresh milk, the serum of which was prepared in the manner already described. These samples of milk were treated with chloro-

[1] U. S. Dept. Agr. Soils. Bul. 19, p. 31, and Bul. 31, pp. 12-16.
[2] U. S. Dept. Agr. An. Ind. Bul. 166, p. 9.

form at the rate of 50 c.c. per 1000 c.c. of milk and the fat removed by means of a centrifugal machine; the removal of fat is necessary since it clogs the pores of the filter. The fat-free milk was then filtered through Pasteur-Chamberland filtering tubes. Analyses were made of the milk and of the serum. We did not determine those constituents present in milk only in traces, such as iron, sulphuric acid, etc.

TABLE I.— CONSTITUENTS OF MILK-SERUM.

CONSTITUENTS.	SAMPLE No. 1.			SAMPLE No. 2.		
	Original milk 100 c.c.	Milk-serum 100 c.c.	Percentage of milk constituents in serum.	Original milk 100 c.c.	Milk-serum 100 c.c.	Percentage of milk constituents in serum.
	Grams.	Grams.	Per ct.	Grams.	Grams.	Per ct.
Sugar......................	......	......	......	5.75	5.75	100.00
Casein...................	3.35	0.00	0.00	3.07	0.00	0.00
Albumin................	0.525	0.369	70.29	0.506	0.188	37.15
Nitrogen in other compounds..................	......	......	......	0.049	0.049	100.00
Citric acid.............	......	......	......	0.237	0.237	100.00
Phosphorus (organic and inorganic).............	0.125	0.067	53.60	......	......	......
Phosphorus (inorganic)...	0.096	0.067	70.00	0.087	0.056	64.40
Calcium................	0.128	0.045	35.16	0.144	0.048	33.33
Magnesium.............	0.012	0.009	75.00	0.013	0.007	53.85
Potassium............ } Sodium.............. }	*0.354	*0.352	99.44	{ 0.120 { 0.055	0.124 0.057	100.00 100.00
Chlorine.............	0.081	0.082	100.00	0.076	0.081	100.00
Ash..................	......	......	......	0.725	0.400	55.17

* As chlorides

A study of the data contained in Table I enables us to show the general relation of the constituents of milk to the constituents of milk-serum. The following form of statement furnishes a clear summary of the facts.

1. Milk constituents in true solution in milk serum:

(a) Sugar.
(b) Citric acid
(c) Potassium.
(d) Sodium.
(e) Chlorine.

2. Milk constituents partly in solution and partly in suspension or colloidal solution:

(a) Albumin.
(b) Inorganic phosphates.
(c) Calcium.
(d) Magnesium.

3. Milk constituents entirely in suspension or colloidal solution:

(a) Fat.
(b) Casein.

The behavior of milk albumin attracts special attention on account of marked lack of regularity in the results obtained. We commonly think of milk albumin as readily and completely soluble in water, and the question is therefore raised as to why a considerable portion of it does not pass through the Pasteur-Chamberland filter. In view of all the facts available, the most probable explanation that has so far suggested itself is that in fresh milk a part of the albumin is held by the adsorbing power of casein. This suggestion is supported by results obtained in the following experiments: Serum was prepared from chloroformed fresh milk treated in different ways. In the first experiment, serum direct from the fresh milk was compared with serum obtained from whey which had been obtained from another portion of the same milk by treatment with rennet-extract. In the second experiment, serum direct from fresh milk was compared with (a) serum obtained from another portion of the same milk after souring, and (b) serum obtained from another portion of the same milk to which some formaldehyde solution had been added. Albumin was determined in each case by boiling after addition of acetic acid, following the details given in the provisional method of the Association of Official Agricultural Chemists. The results of the experiments are given below.

	Albumin per 100 c.c. Grams.	Albumin of milk recovered in serum. Per ct.
FIRST EXPERIMENT.		
Fresh milk..................	0.312	......
Serum from fresh milk.........	0.143	45.83
Serum from whey.............	0.187	59.94
SECOND EXPERIMENT.		
Fresh milk..................	0.266	......
Serum from fresh milk.........	0.148	55.64
Serum from sour milk.........	0.253	95.11
Serum from milk plus formaldehyde......................	0.245	92.21

In the first experiment it is seen that when casein is precipitated by rennet solution the curd (the precipitated casein or paracasein) carries down part of the albumin with it; the amount thus carried down is approximately equal in this case to that retained along with the casein on the external surface of the Pasteur-Chamberland filtering tube, when whole milk is filtered through such a filter.

In the second experiment we see that when the casein is precipitated with acid, as in the case of natural souring, the adsorbing action of the casein is practically prevented and little or no albumin is carried down with it. In the case of the addition of formaldehyde to milk, the adsorbing power of casein is greatly diminished, probably due to the chemical reaction between casein and formaldehyde.

PROPERTIES AND COMPOSITION OF PORTION OF MILK IN SUSPENSION OR COLLOIDAL SOLUTION.

Some of the constituents of milk are suspended in the form of solid particles in such an extremely fine state of division that they pass through the pores of filter paper and they do not settle as a sediment on standing, but remain permanently afloat. They can not be seen except by ultra-microscopic methods. When substances are in such a condition, they are said to form a colloidal solution. In passing milk through the Pasteur-Chamberland filtering tube, the constituents in suspension as solid particles, and in colloidal solution, are retained in a solid mass on the outside of the tube and can therefore be readily obtained for study.

(1) *Appearance.*— When prepared by the method of filtration previously described, the insoluble portion of milk collecting on the outside of the filtering tube is grayish to greenish white in color, of a glistening, slime-like appearance and gelatinous consistency. When dried without purification by treatment with alcohol, etc., it resembles in appearance dried white of egg.

(2) *Behavior with water.*—The deposit of insoluble milk-constituents on the outside of the filtering tube, when removed and shaken vigorously in a flask with distilled water, goes into suspension and the mixture has the opaque, white appearance of the original milk. The deposit is, of course, more or less mixed with adhering soluble constituents but can be readily purified by shaking with distilled water and filtering several times. The purified material goes readily into suspension on shaking with water and, if treated with a preservative, will remain indefinitely without change other than the separation of fat-globules. It has been held by some that the citrates of milk perform the function of holding the insoluble phosphates in suspension, but this is not supported by the behavior of the insoluble portion shown in our experiments.

(3) *Reaction.*— A suspension of the insoluble constituents of milk, prepared in the manner described above, is neutral to phenolphthalein. We purified the deposit made from 1000 c.c. of milk, made a suspension of it in water, and, after the addition of 10 c.c. of neutral solution of potassium oxalate, it was found to require only 0.5 c.c. of $\frac{N}{10}$ solution of sodium hydroxide to make it neutral to phenolphthalein: We interpret this to mean that there are no tri-basic (alkaline) phosphates in milk or in the serum, because the serum,

since it is acid, can contain none, and the insoluble portion, being neutral, can therefore contain none.

(4) *Relation of inorganic constituents to casein in milk.*— Without going into a detailed discussion of the history of the different views held by different investigators, it is sufficient for our purpose to state that three general views have been put forward in regard to the relation of inorganic constituents to casein in milk: (1) That milk-casein is combined with calcium (about 1.07 per ct.) to form a salt, calcium caseinate (which is neutral to litmus and acid to phenol-phthalein); (2) that casein is chemically combined directly with calcium phosphate; (3) that casein is a double compound consisting of calcium caseinate combined with calcium phosphate.

We have attempted to learn what is the true condition of casein in milk in relation to inorganic constituents, whether it is in combination with calcium alone or with some other inorganic base in addition and also whether milk-casein is an acid salt or a neutral salt and, further, whether the insoluble phosphates are in combination with casein or not.

In studying this problem, we will first give results of work done with 16 samples of fresh milk from 13 individual cows. Determinations were made of (a) casein, (b) total phosphorus, (c) soluble phosphorus, (d) insoluble phosphorus (b minus c), (e) insoluble organic phosphorus (0.71 per ct. of the casein), (f) insoluble inorganic phosphorus (d minus e), (g) total calcium, (h) soluble calcium,

TABLE II.— AMOUNTS OF PROTEINS, CASEIN, AND PHOSPHORUS IN MILK.

Cow No.	Stage of lactation	Total proteins.	Casein.	PHOSPHORUS.					Ratio of organic to insoluble inorganic phosphorus.
				Total.	Soluble.	INSOLUBLE.			
						Total.	Organic (in casein).	Inorganic (phosphates)	
(1)	(2)	(3)	(4)	(5)	(6)	(7)	(8)	(9)	(10)
		Per ct.	*Per ct.*	*Per ct.*	*Per ct.*	*Per ct.*	*Per ct.*	*Per ct.*	Org. In. P. P.
1	3 days..	4.35	3.48	0.1272	0.0818	0.0454	0.0247	0.0207	1 : 0.83
2	1 mo...	3.31	2.73	0.1150	0.0595	0.0555	0.0194	0.0361	1 : 1.86
3	1 " ...	3.53	2.78	0.1010	0.0494	0.0516	0.0197	0.0319	1 : 1.62
3	11 " ...	4.91	4.09	0.1111	0.0536	0.0575	0.0290	0.0285	1 : 0.98
4	3 " ...	3.93	3.09	0.1278	0.0563	0.0715	0.0219	0.0496	1 : 2.26
5	3 " ...	3.45	2.88	0.0943	0.0475	0.0468	0.0204	0.0264	1 : 1.29
5	7 " ...	3.45	2.70	0.0870	0.0334	0.0536	0.0192	0.0344	1 : 1.79
6	5 " ...	4.05	2.92	0.1008	0.0356	0.0652	0.0207	0.0445	1 : 2.15
7	6 " ...	4.07	3.40	0.1063	0.0548	0.0515	0.0241	0.0274	1 : 1.14
7	10 " ...	4.80	3.56	0.1010	0.0340	0.0670	0.0253	0.0417	1 : 1.65
8	7 " ...	4.39	3.58	0.1157	0.0550	0.0607	0.0254	0.0353	1 : 1.39
9	8 " ...	4.33	3.47	0.1036	0.0364	0.0672	0.0246	0.0426	1 : 1.73
10	9 " ...	3.65	3.10	0.1097	0.0610	0.0487	0.0220	0.0267	1 : 1.22
11	10 " ...	4.17	3.36	0.1090	0.0434	0.0656	0.0239	0.0417	1 : 1.74
12	11 " ...	4.35	3.14	0.1060	0.0286	0.0774	0.0223	0.0551	1 : 2.47
13	12 " ...	5.71	4.97	0.1310	0.0442	0.0868	0.0353	0.0515	1 : 1.46

(i) insoluble calcium (g minus h), (j) total magnesium, (k) soluble magnesium, (l) insoluble magnesium (j minus k). The determinations of total phosphorus, total calcium and total magnesium were made with the normal or whole milk, while those of soluble phosphorus, soluble calcium and soluble magnesium were made with the serum obtained by filtering through Pasteur-Chamberland filtering tubes in the manner already described. The amount of organic phosphorus was found[3] by multiplying the percentage of casein by 0.0071. For convenience of reference, the analytical data are arranged in two tables, II and III.

The data in Table II afford a basis for ascertaining the quantitative relation between casein and the phosphates. If casein is chemically combined with phosphates in milk, there should be a fairly definite and uniform relation between these constituents in the insoluble portion of milk, or, stated in another way, the organic phosphorus of casein should show a somewhat uniform ratio to the insoluble inorganic or phosphate phosphorus. In column 10 of Table II are given the results of calculations based on our data, which show the amount of insoluble inorganic phosphorus for one part of organic (casein) phosphorus. It is seen that the ratio varies between the wide limits of 1:0.82 and 1:2.47. Even in the case of milk from the same animal at different stages of lactation, the proportional amount of inorganic phosphorus varies widely, as from 0.98 to 1.62 with cow No. 3, from 1.29 to 1.79 with cow No. 5, and from 1.14 to 1.65 with cow No. 7. The only conclusion furnished by these results is that there is no evidence of chemical combination between the casein and the phosphates of milk. Additional evidence in confirmation of the foregoing statement will be furnished later in connection with the discussion of another phase of the subject.

Another interesting point connected with insoluble phosphates and casein in milk is as to the exact compound of calcium phosphate and of calcium caseinate existing in the milk. Söldner's inferential statement that milk-casein is neutral calcium caseinate (containing about 1.07 per ct. of calcium), has been generally accepted, not so much because of positive proof but because of absence of any proof to the contrary. Regarding the form of the compound in which phosphates exist in milk, all three forms (mono-, di-, and tri-basic phosphates) have been thought to be present. The insoluble phosphates have been regarded as a mixture of di- and tri-calcium phosphates.

Bearing on this question, we present data embodied in the following tables, III and IV.

[3] Bosworth and Van Slyke. N. Y. Agrl. Expt. Sta. Tech. Bul. 37, and Jour. *Biol. Chem.*, 19:67.

TABLE III.— AMOUNTS OF CALCIUM AND MAGNESIUM IN INSOLUBLE PORTION OF MILK.

Cow No.	Stage of lactation.	CALCIUM.			MAGNESIUM.		
		Total.	Soluble.	Insoluble.	Total.	Soluble.	Insoluble.
		(11)	(12)	(13)	(14)	(15)	(16)
		Per ct.	Per ct.	Per ct.	Per ct.	Per ct.	Per ct.
1	3 days..	0.1607	0.0734	0.0873	0.0156	0.0142	0.0013
2	1 mo...	0.1381	0.0511	0.0870	0.0136	0.0117	0.0019
3	1 "	0.1362	0.0544	0.0818	0.0180	0.0142	0.0038
3	11 "	0.1559	0.0534	0.1025	0.0170	0.0156	0.0014
4	3 "	0.1483	0.0343	0.1140	0.0184	0.0128	0.0056
5	3 "	0.1396	0.0531	0.0865	0.0156	0.0124	0.0032
5	7 "	0.1256	0.0454	0.0802	0.0147	0.0134	0.0013
6	5 "	0.1413	0.0373	0.1040	0.0160	0.0127	0.0033
7	6 "	0.1464	0.0526	0.0938	0.0144	0.0121	0.0023
7	10 "	0.1523	0.0450	0.1073	0.0177	0.0127	0.0050
8	7 "	0.1506	0.0439	0.1062	0.0153	0.0118	0.0035
9	8 "	0.1503	0.0440	0.1063	0.0171	0.0126	0.0045
10	9 "	0.1410	0.0543	0.0867	0.0168	0.0141	0.0027
11	10 "	0.1379	0.0357	0.1022	0.0168	0.0119	0.0049
12	11 "	0.1659	0.0414	0.1245	0.0191	0.0123	0.0068
13	12 "	0.2167	0.0669	0.1498	0.0236	0.0163	0.0073

The data in Table IV are derived by calculation from the figures given in Tables II and III for the purpose of reducing them to a uniform basis that permits us to make comparison more easily.

TABLE IV.— AMOUNTS OF ACIDS AND BASES EXPRESSED AS GRAM EQUIVALENTS.

Cow No.	Casein as gram equivalents of octavalent acid.	Insoluble inorganic phosphates as gram equivalents of di-basic acid.	Sum of gram equivalents of casein and phosphates.	Insoluble calcium as gram equivalents.	Insoluble magnesium as gram equivalents.	Sum of the insoluble calcium and magnesium as gram equivalents.	Excess of insoluble base (+) or acid (—) as gram equivalents.
(1)	(2)	(3)	(4)	(5)	(6)	(7)	(8)
1	31.3×10^{-4}	13.1×10^{-4}	44.4×10^{-4}	43.6×10^{-4}	1.1×10^{-4}	44.7×10^{-4}	$+0.3 \times 10^{-4}$
2	24.5 "	23.1 "	47.6 "	43.5 "	1.6 "	45.1 "	—2.5 "
3	25.0 "	20.5 "	45.5 "	40.9 "	3.1 "	44.0 "	—1.5 "
3	36.8 "	18.1 "	54.9 "	51.2 "	1.2 "	52.4 "	—2.5 "
4	27.8 "	32.0 "	59.8 "	57.0 "	4.6 "	61.6 "	+1.8 "
5	25.9 "	16.8 "	42.7 "	43.2 "	2.7 "	45.9 "	+3.2 "
5	24.3 "	22.2 "	46.5 "	40.1 "	1.0 "	41.1 "	—5.4 "
6	26.3 "	28.7 "	55.0 "	52.0 "	2.7 "	54.7 "	—0.3 "
7	30.6 "	17.4 "	48.0 "	46.9 "	1.9 "	48.8 "	+0.8 "
7	32.0 "	26.9 "	58.9 "	53.7 "	4.2 "	57.9 "	—1.0 "
8	32.2 "	22.3 "	54.5 "	53.4 "	2.9 "	56.3 "	+1.8 "
9	31.1 "	27.5 "	58.6 "	53.2 "	3.8 "	57.0 "	—1.6 "
10	27.9 "	17.0 "	44.9 "	43.4 "	2.2 "	45.6 "	+0.7 "
11	30.2 "	27.0 "	57.2 "	51.1 "	4.1 "	55.2 "	—2.0 "
12	28.3 "	35.5 "	63.8 "	62.3 "	5.7 "	68.0 "	+4.2 "
13	44.7 "	32.9 "	77.6 "	74.9 "	6.1 "	81.0 "	+3.4 "

In our previous work we have shown that 1 gram of uncombined casein combines with 9×10^{-4} gram equivalents of calcium to form a salt that is neutral to phenolphthalein.[4] In column 2 of Table IV we make use of this fact in calculating the acid equivalents of the casein as found in each sample. In column 3 of the same table, we calculate the acid equivalents of the insoluble inorganic phosphorus in each sample of milk (regarding phosphoric acid as a divalent acid and $CaHPO_4$ neutral to phenolphthalein). In column 4 are shown the sums obtained by adding the figures in columns 2 and 3 in case of each sample of milk. In columns 5 and 6 are given the combining equivalents of calcium and magnesium and in column 7 their sum for each sample of milk. If now we compare, in case of each milk, the figures contained in column 4 with those contained in column 7, we notice that they are in close agreement, the differences being shown in column 8. This agreement means that the quantitative relation between the bases (calcium and magnesium) and the acids (casein and phosphoric acid) is that required, theoretically, to give di-calcium phosphate with a trace of di-magnesium phosphate and calcium caseinate neutral to phenolphthalein, in which casein is combined with 8 equivalents of calcium (casein Ca_4). However, the same analytical figures can with equal correctness be interpreted to prove that the compounds are present as acid caseinate and tri-calcium phosphate.

In order to decide which of these sets of compounds is present in milk, we have tried to make a separation of the casein and insoluble phosphates. The above results, it will be remembered, are obtained by difference, the milk and serum being analyzed and the composition of the insoluble portion being determined by subtracting the latter results from the former. It seemed desirable to separate milk in large amounts so as to obtain the insoluble portion in quantity sufficient to purify and analyze. This was done in the following manner, several experiments being made. In the first experiment, 400 pounds of milk was run through a centrifugal cream separator 18 times and the deposit ("separator slime") collecting on the walls of the bowl was removed after the 1st, the 6th, the 12th and the 18th run. Each of these deposits was placed in a mortar and triturated with small amounts of 95 per ct. alcohol with the gradual addition of more alcohol. A point is reached when the whole mass becomes jelly-like, after which the addition of more alcohol causes the formation of a fine flocculent precipitate. (Care must be taken not to add the alcohol too rapidly, because then there is apt to be formed a tough, leathery mass, which can not be handled.) The precipitate is allowed to settle and, after decanting the supernatant liquid, is triturated with several successive portions of 95 per ct. alcohol, 99 per ct. alcohol, and finally ether. It is then dried

[4] N. Y. Agrl. Expt. Sta. Tech. Bul. No. 26, p. 12.

at 60° C. for a few hours, after which the drying is completed in a vacuum over sulphuric acid. The analytical results are given in the table following:

TABLE V.— COMPOSITION OF INSOLUBLE PORTION ("SEPARATOR SLIME") OF MILK.

Sample of deposit taken.	Casein.	Ash.	Total phosphorus.	Phosphorus in casein.	Phosphorus as phosphate.	Calcium.	Ratio of organic to insoluble inorganic phosphorus.
	Per ct.	Per ct.	Per ct.	Per ct.	Per ct.	Per ct.	Org.P.: In.P.
After 1st run...	86.31	10.43	2.182	0.621	1.561	3.386	1 : 2.51
After 6th run...	90.07	9.35	1.950	0.649	1.301	3.246	1 : 2.00
After 12th run...	90.84	9.53	2.011	0.645	1.366	3.343	1 : 2.11
After 18th run...	91.98	9.62	2.023	0.662	1.361	3.223	1 : 2.06

The figures in Table V, obtained by direct analysis of the insoluble deposit or "separator slime," show a striking agreement with the results obtained by the indirect method, which is brought out more clearly by expressing the above figures in the form of gram equivalents, as follows:

TABLE VI.— AMOUNTS OF ACIDS AND BASES EXPRESSED AS GRAM EQUIVALENTS.

Sample of deposit taken.	Casein as gram equivalents of acid.	Phosphates as gram equivalents of di-basic acid.	Sum of gram equivalents of casein and phosphates.	Gram equivalents of calcium.
After 1st run...............	77.7×10^{-3}	100.7×10^{-3}	178.4×10^{-3}	169.3×10^{-3}
After 6th run...............	81.1 "	82.9 "	164.0 "	162.3 "
After 12th run...............	80.8 "	88.1 "	168.9 "	167.2 "
After 18th run...............	82.8 "	87.8 "	170.6 "	161.2 "

The high percentage of inorganic phosphorus in the deposit from the first run indicates that the phosphates are heavier than the caseinates and could be separated from them if a certain speed were used in running the separator. This point is further shown by the following experiments: In the first experiment, the bowl of a cream separator was filled with fat-free milk (about 1,000 c.c.) and was whirled for two hours, at a speed of 5,000 revolutions per minute, when the milk was taken out and the "separator slime" which had collected on the bowl was removed and treated with alcohol and ether in the manner already described. The same milk was returned to the separator bowl and again whirled for two hours, when the deposit was again removed and treated as before. When removed the second time, that is, after four hours of whirling, the milk was nearly as clear as whey, most of the suspended phosphates and casein having been deposited on the walls of the bowl during the whirling. The results of analysis of the " separator slime " deposited after each two hours of whirling are given in the table following:

TABLE VII.— COMPOSITION OF INSOLUBLE PORTION OF MILK DEPOSITED AT DIFFERENT INTERVALS.

" Slime " formed by whirling two 2-hour periods.	Casein.	Ash.	Total phos- phorus.	Phos- phorus in casein.	Phos- phorus as phosphate.	Calcium.	Ratio of organic to insoluble inorganic phosphorus.
	Per ct.	Per ct.	Per ct.	Per ct.	Per ct.	Per ct	Org. P.:In. P.
1st 2 hours.....	90.68	9.32	1.909	0.653	1.256	3.090	1 : 1.92
2nd 2 hours....	91.12	8.88	1.437	0.656	0.781	2.691	1 : 1.19

These results show that two-thirds of the insoluble inorganic phosphorus was removed during the first two hours of whirling, again indicating that the phosphates are heavier than the casein. The ratio of casein to phosphates is here also shown to be wholly irregular, indicating no definite combination.

Expressing the data in Table VII in the form of gram equivalents, we have the figures contained in the following table:

TABLE VIII.— AMOUNTS OF ACIDS AND BASES EXPRESSED AS GRAM EQUIVALENTS.

	Casein as gram equivalents of acid.	Phosphates as gram equivalents of di-basic acid.	Sum of gram equivalents of casein and phosphates.	Gram equivalents of calcium.
1st deposit..................	81.6×10^{-3}	81.1×10^{-3}	162.7×10^{-3}	154.5×10^{-3}
2nd deposit..........	82.0×10^{-3}	50.4×10^{-3}	132.4×10^{-3}	134.6×10^{-3}

An examination of these figures shows that there is the same balance between the acids (casein and phosphoric acid) and the bases (calcium and magnesium) in the two separate deposits, even when the inorganic phosphorus is as unevenly distributed between them, which furnishes proof for two points: (1st) The inorganic phosphorus must be in the form of neutral calcium phosphate ($CaHPO_4$), for otherwise the balance between bases and acids would be altered, acid calcium phosphate ($CaH_4P_2O_8$) giving an excess of acid and tri-calcium phosphate ($Ca_3P_2O_8$) an excess of base in the " slime " deposited in the first whirling. (2nd) If the phosphates were in combination with the casein, we should expect to find the ratio between the organic phosphorus and the inorganic phosphorus the same in both deposits, but, instead of uniformity, we find the ratio showing so wide a variation as 1:1.92 and 1:1.19 in the two cases.

In the second experiment, further evidence is furnished, showing that neutral calcium phosphate ($CaHPO_4$) is a normal constituent of milk. Four 500-c.c. bottles were filled with separator skim-milk to which some formaldehyde had been added, and, after standing at room temperature for 4 days, were whirled in a Bausch and Lomb precision centrifugal machine for 30 minutes at a speed of 1,200 revolutions per minute. A sediment was deposited, which after purification by treatment with alcohol and ether, as previously described, weighed 0.4 gram. Analysis of this gave the following results: Casein, 20.78 per ct.; total phosphorus, 18.38 per ct.; phosphorus combined with casein, 0.15 per ct.; phosphorus combined as phosphates, 18.23 per ct.; calcium, 22.79 per ct.; ratio of organic phosphorus to inorganic phosphorus, 1:121; casein as gram equivalents of acid, 18.7×10^{-3}; phosphates as gram equivalents of di-basic acid, $1175. \times 10^{-3}$; sum of casein and phosphates as gram equivalents of acid, $1194. \times 10^{-3}$; gram equivalents of calcium, $1140. \times 10^{-3}$.

In these figures, we again find the same balance between bases and acids, which can mean only that the phosphate compound deposited is di-calcium phosphate ($CaHPO_4$). The degree of centrifugal force developed was sufficient to throw out a relatively large amount of di-calcium phosphate but not powerful enough to throw out very much casein, thus serving as a means of effecting a nearly complete separation of these two constituents.

Babcock [5] whirled skim-milk in a separator for several hours, removing portions from time to time for analysis and finally determining the amounts of casein, calcium and phosphorus in the deposited " slime." While the experiments were preliminary in character and the results not sufficient to base permanent conclusions on, they tended to show that the casein and phosphates were not in combination. From the analytical results showing the relation of calcium to phosphorus, the conclusion was drawn that tri-calcium phosphate is the compound present in milk. The figures for calcium and phosphorus were based upon the total amounts contained in the deposit and no allowance was made for the calcium in combination with casein and the phosphorus of the casein. This fact accounts for the difference between the results presented by him and the conclusions reached by us. A recalculation of his data, after deducting the amounts of calcium and phosphorus combined with casein, gives figures that correspond to the composition of $CaHPO_4$ and not $Ca_3P_2O_8$, thus confirming the results of our work.

ACIDITY OF MILK AND MILK-SERUM.

Both fresh milk and the serum from fresh milk show a slight acid reaction to phenolphthalein. This has been believed to be due

[5] Wis. Agrl. Expt. Sta. 12th An. Rept., p. 93.

to casein or acid phosphates in the milk or to both. The fact that fresh milk and its serum are strongly alkaline to methyl orange. indicates that the acidity is due to acid phosphates, though it does not necessarily show that acid caseinates are not also responsible for some of the acidity. The results of our work given in the preceding pages furnish aid in determining to what compounds in milk the acid reaction to phenolphthalein is due.

A 1,000 c.c. sample of milk was obtained from each of eight cows immediately after milking and chloroform (50 c.c.) was added to this at once. The acidity of the milk and of the milk-serum was determined after treatment with neutral potassium oxalate according to the method of Van Slyke and Bosworth.[6] The results are given below.

TABLE IX.— ACIDITY OF MILK AND MILK-SERUM.

NUMBER OF SAMPLE.	NUMBER OF C.C. OF $\frac{N}{10}$ ALKALI REQUIRED TO NEUTRALIZE 100 C.C. OF —	
	Milk.	Milk-serum.
1	4.8	5.0
2	6.2	6.2
3	4.2	4.2
4	6.0	5.8
5	6.4	6.4
6	4.4	4.4
7	7.0	6.8
8	6.6	6.4

These figures show that the acidity of fresh milk is the same as that of its serum which means that the constituents of the milk causing acidity are soluble constituents contained in the serum. Since the serum contains phosphates in amounts sufficient to furnish two to four times as much acid phosphates as is required to account for the acidity, and since, moreover, no other acid constituents of the milk-serum are present in more than minute quantities and are wholly insufficient to cause the observed degree of acidity, it appears a reasonable conclusion that the acidity of fresh milk is due to soluble acid phosphates. This conclusion is further strengthened by the results given in the preceding pages which go to show conclusively that the insoluble constituents of fresh milk are neutral in reaction, consisting largely or wholly of neutral calcium caseinate (casein Ca_4), neutral di-calcium phosphate ($CaHPO_4$) and fat.

[6] N. Y. Agrl. Expt. Sta. Tech. Bul. No. 37, p. 5.

COMPOUNDS OF MILK.

It is difficult to learn what are the individual forms or compounds in which the salts exist in milk. Attempts have been made to determine this by inferences based on analytical results. In view of the data presented in the preceding pages, taken together with many other analytical data worked out by us, we suggest the following statement as representing in some respects more closely than previous ones, facts corresponding to our present knowledge of the principal constituents of milk. The amounts are based on milk of average composition.

Fat...	3.90 per ct.
Milk-sugar.................................	4.90 "
Proteins combined with calcium..............	3.20 "
Di-calcium phosphate ($CaHPO_4$)	0.175
Calcium chloride ($CaCl_2$)....................	0.119 "
Mono-magnesium phosphate ($MgH_4P_2O_8$)......	0.103
Sodium citrate ($Na_3C_6H_5O_7$).................	0.222
Potassium citrate ($K_3C_6H_5O_7$)...............	0.052
Di-potassium phosphate (K_2HPO_4)...........	0.230 "
Total solids............................	12.901 per ct.

8+b

TECHNICAL BULLETIN No. 40.   JANUARY, 1915.

New York Agricultural Experiment Station.

GENEVA, N. Y.

CONCERNING THE ORGANIC PHOSPHORUS COMPOUND OF WHEAT BRAN AND THE HYDROLYSIS OF PHYTIN.

I.   INOSITE TRIPHOSPHORIC ACID IN WHEAT BRAN.
(Eleventh paper on Phytin.)

II. HYDROLYSIS OF PHYTIN BY THE ENZYME PHYTASE.
(Twelfth paper on Phytin.)

III. HYDROLYSIS OF THE ORGANIC PHOSPHORUS COMPOUND OF WHEAT
BRAN BY THE ENZYME PHYTASE.
(Thirteenth paper on Phytin.)

IV. PHYTIN IN WHEAT BRAN.
(Fourteenth paper on Phytin.)

R. J. ANDERSON.

PUBLISHED BY THE DEPARTMENT OF AGRICULTURE

CONCERNING THE ORGANIC PHOSPHORUS COMPOUND OF WHEAT BRAN AND THE HYDROLYSIS OF PHYTIN.

R. J. ANDERSON

SUMMARY.

This bulletin contains reports of investigations concerning: (I) the nature and composition of the principal organic phosphoric acid isolated from the 0.2 per ct. hydrochloric acid extract of wheat bran, (II) the products formed from phytin by the action of the enzyme phytase contained in wheat bran, (III) the hydrolysis of the organic phosphorus compound of wheat bran in different solvents and (IV) the nature and composition of the organic phosphorus compound of wheat bran when isolated from solvents which destroy the enzyme phytase.

It has been shown in previous reports that the organic phosphorus compounds isolated from 0.2 per ct. hydrochloric acid extracts of wheat bran differ in composition from phytin and phytic acid or inosite hexaphosphoric acid. (Technical Bulletin No. 22 of this Station.)

It has also been shown that the above substance is not a homogeneous compound but that it can be separated into several fractions which differ in composition. (Technical Bulletin No. 36 of this Station.)

I. The first part of this bulletin describes the further separation of these compounds and the isolation of a new organic phosphoric acid, inosite triphosphoric acid, as a crystalline strychnine salt from the water-insoluble portion of the acid barium salts.

The neutral barium inosite triphosphate, $C_6H_9O_{15}P_3Ba_3$, was prepared from the crystalline strychnine salt. It was a white amorphous powder.

The free inosite triphosphoric acid $C_6H_{15}O_{15}P_3$, was prepared from the barium salt and it was obtained as a non-crystallizable syrup.

The reactions of inosite triphosphoric acid differ in several particulars from those of phytic acid or inosite hexaphosphoric acid but like the latter it decomposes when heated in a sealed tube with dilute sulphuric acid into inosite and phosphoric acid.

II. The chief products of the hydrolysis of phytin by the phytase in wheat bran are inorganic phosphoric acid and certain intermediate compounds apparently consisting of inosite tri-, di- and monophosphoric acids. These intermediate substances are identical with the compounds which we have previously isolated from 0.2 per ct. hydrochloric acid extracts of wheat bran.

A portion of the phytin was completely hydrolyzed by the action of the enzyme into phosphoric acid and inosite because the solution was found to contain some free inosite.

All of the phytin was partially hydrolyzed since the final reaction mixture did not contain any unchanged inosite hexaphosphoric acid.

III. The results herein reported amplify and confirm the experiments of Suzuki, Yoshimura and Takaishi and of Plimmer concerning the presence of the enzyme "phytase" in wheat bran which is capable of hydrolyzing phytin with the production of inorganic phosphoric acid.

The maximum activity of the enzyme has been shown to occur in the presence of o.1 per ct. hydrochloric acid and o.2 per ct. acetic acid. With increasing concentration of the hydrochloric acid the activity rapidly diminishes and with o.5 per ct. hydrochloric acid there is practically no hydrolysis of the organic phosphorus.

The enzyme is destroyed by boiling water and by boiling o.2 per ct. hydrochloric acid. It is also destroyed by a short exposure to o.5 per ct. hydrochloric acid and to o.25 per ct. ammonia.

It is shown that wheat bran normally contains about o.1 per ct. of inorganic phosphorus, which is equal to about 11 per ct. of the total soluble phosphorus.

IV. By digesting wheat bran in 1.0 per ct. hydrochloric acid, which is sufficiently strong to destroy the enzyme phytase, it is possible to isolate from the extract crystalline barium salts of the following composition:

$$C_6H_{12}O_{24}P_6Ba_3 + 8H_2O \text{ and } (C_6H_{11}O_{24}P_6)_2Ba_7 + 14H_2O$$

These salts are identical with the tribarium phytate and heptabarium phytate obtained from oats, corn, cottonseed meal and commercial phytin. All of these materials contain therefore the same organic phosphorus compound, viz., phytic acid or inosite hexaphosphoric acid, $C_6H_{18}O_{24}P_6$.

I. INOSITE TRIPHOSPHORIC ACID IN WHEAT BRAN.[*]

(Eleventh paper on Phytin.)

INTRODUCTION.

The only definitely homogeneous organic phosphoric acid ever isolated from wheat bran, so far as we are aware, is the crystalline inosite monophosphate which we described in a previous paper.[1] This substance differs from other inosite phosphoric acids in that it crystallizes readily and particularly in that its barium salt is very soluble in water, i. e., it is not precipitable with barium hydroxide.

---

[*] The experimental work herein reported was carried out mainly in the First Chemical Institute of the University of Berlin, Berlin, Germany.

[1] Jour. Biol. Chem. 18: 441, 1914; and N. Y. Agr. Exp. Sta. Tech. Bull. 36, 1914.

Wheat bran, however, contains other organic phosphoric acids which are precipitated by barium hydroxide. It is evident from the work which we last reported [2] that this water-insoluble barium salt is not a homogeneous substance but that it is a mixture of various organic phosphoric acids. Patten and Hart [3] isolated an acid from this mixture of the water-insoluble barium salts, which they believed to be phytic acid. We have shown,[4] however, by the analyses of numerous preparations, freed from inorganic phosphate, that the above substance differs entirely in composition from phytic acid or salts of this acid. Although we were able to separate this substance into various fractions which all differed in composition, it was impossible to obtain definitely homogeneous compounds by the method employed. The separation was the more difficult since neither the barium salts nor other salts with inorganic bases crystallized.

In our previous work on wheat bran we had always used the crude bran ordinarily sold in the market for cattle feed. This product was not a pure bran but contained various impurities. It is not unlikely that these impurities also contained organic phosphoric acids which would render the separation of pure compounds from the final mixture more difficult.

In the work reported in this paper we have used a perfectly pure wheat bran which was especially prepared from winter wheat for us in a local mill.

From this pure bran the organic phosphoric acids were isolated as barium salts in the usual way (see Experimental Part). The acid barium salt finally obtained was shaken up with cold water in which about one-half of the total substance dissolved. *The examination of the water-insoluble portion forms the subject of this paper.* We hope, later on, to be able to separate the *water-soluble* substance, also, into its constituents.

The portion insoluble in water, as was shown in our last report,[5] contains a higher percentage of phosphorus and a lower percentage of carbon than the water-soluble substance.

Since the barium salt of the above organic phosphoric acid as well as salts with other inorganic bases do not crystallize we tried to prepare salts with organic bases in the hope of obtaining crystalline compounds. It was found, however, that salts with phenylhydrazin, hydrazin hydrate, aniline, pyridine, etc., showed no more tendency to crystallize than those with inorganic bases; and the brucine salt is too soluble to crystallize well.

[2] *Jour. Biol. Chem.* 18:425, 1914; and N.Y. Agr. Exp. Sta. Tech. Bull. 36, 1914.
[3] *Amer. Chem. Jour.* 31:566, 1904.
[4] *Jour. Biol. Chem.* 12:447, 1912; 18:425, 1914; and N. Y. Agr. Exp. Sta. Tech. Bull. 22, 1912, and Tech. Bull. 36, 1914.
[5] *Jour. Biol. Chem.* 18:425, 1914; and N. Y. Agric. Exp. Sta. Tech. Bull. 36, 1914.

It was found finally that strychnine gave a readily crystalline salt with the above acid. This strychnine salt separates from the hot aqueous solution, on cooling, in long needle-shaped crystals or sometimes in large thin plates, depending upon the concentration. The salt was then purified by recrystallizing several times from water.

This substance had practically the same composition and melting point as the salts described by Clarke [6] which he had prepared from wild Indian mustard.

However, it must be noted that the molecular weight of such strychnine salts is so high that it is impossible accurately to determine the composition of the acid from the analysis of these compounds. We have used the strychnine salts therefore merely as a means of purification. After recrystallizing the salt several times the strychnine was removed and the acid again precipitated with barium hydroxide.

The substance was then analyzed and the composition agreed very closely with a neutral barium salt of inosite triphosphate, $C_6H_9O_{15}P_3Ba_3$. An acid barium salt was prepared by precipitating the hydrochloric acid solution of the neutral salt with alcohol. On analysis results were obtained which agreed with a compound of the composition $C_{18}H_{35}O_{45}P_9Ba_5$, which apparently represents three molecules of monobarium inosite triphosphate joined by two atoms of barium.

The free acid was prepared from the barium salt and analyzed. The results agreed closely with the calculated composition of inosite triphosphoric acid, $C_6H_{15}O_{15}P_3$.

Unfortunately the barium salts were obtained only as amorphous or granular powders and the free acid itself was a non-crystallizable syrup. We believe nevertheless that the substance represents a pure compound because it had been separated previously as a crystalline strychnine salt which appeared perfectly homogeneous.

On cleavage with dilute sulphuric acid in a sealed tube at a temperature of 150° inosite triphosphate decomposes into inosite and phosphoric acid.

EXPERIMENTAL PART.

ISOLATION OF THE SUBSTANCE FROM WHEAT BRAN AS A BARIUM SALT.

The wheat bran was digested over night in 0.2 per ct. hydrochloric acid. It was then strained through cheesecloth and filtered. The free acid in the extract was nearly neutralized by adding a dilute solution of barium hydroxide until a slight permanent precipitate remained. A concentrated solution of barium chloride was added and the precipitate allowed to settle over night. The supernatant liquid was syphoned off and the residue centrifugalized.

[6] *Jour. Chem. Soc.* 105: 535, 1914.

The precipitate was finally brought upon a Büchner funnel and freed as much as possible from the mother-liquor and then washed in 30 per ct. alcohol. For further purification the substance was precipitated alternately six times from about 1 per ct. hydrochloric acid with barium hydroxide (Kahlbaum, alkali free) and six times with about an equal volume of alcohol.

After this treatment the substance was obtained as a snow-white, amorphous powder. It was free from inorganic phosphate, and bases other than barium could not be detected. But it still contained oxalates from which it was freed in the manner described in a former paper.[7] The substance was finally precipitated from about 1 per ct. hydrochloric acid with alcohol, filtered and washed free of chlorides with dilute alcohol and then in alcohol and ether and dried in vacuum over sulphuric acid. The dry preparation weighed about 66 grams.

This crude acid salt was rubbed up in a mortar with about 500 c.c. of cold water, allowed to stand a few hours and then filtered and washed with water, alcohol and ether and dried in vacuum over sulphuric acid. The water-insoluble residue weighed about 30 grams.

The water-soluble portion contained in the filtrate was precipitated with barium hydroxide and reserved for a future investigation.

PREPARATION OF THE STRYCHNINE SALT.

The water-insoluble portion of the barium salt mentioned above was suspended in water and the barium precipitated with slight excess of dilute sulphuric acid. The barium sulphate was removed and the filtrate precipitated with excess of copper acetate. The copper precipitate was filtered and washed with water until the washings gave no reaction with barium chloride. It was then suspended in water and the copper precipitated with hydrogen sulphide and the copper sulphide filtered off.

The solution containing the free organic phosphoric acid was diluted to about 2 liters with water and heated on the water-bath and 44 grams of powdered strychnine added. After heating for a few minutes the strychnine was dissolved. The solution was filtered and concentrated in vacuum at a temperature of 45° to 50° to about one-half of the volume. The strychnine salt soon began to separate in long needle-shaped crystals. After standing in the ice chest over night the crystals were filtered off and washed in ice-cold water and finally in absolute alcohol and ether and dried in the air. Yield 45.8 grams.

For further purification the substance was twice recrystallized from hot water. It was then obtained in pure white needle-shaped crystals which looked perfectly homogeneous.

From more concentrated solutions the substance sometimes separates in the form of colorless plates which differ from the needle-

7 *Jour. Biol. Chem.* 18:425, 1914; and N. Y. Agrl. Expt. Sta. Tech. Bull. 36, 1914.

shaped crystals in that they contain about 4 per ct. more of water of crystallization.

The strychnine salt has no sharp or definite melting point. Heated in a capillary tube it softens at about 200° C. but it does not melt completely even at a much higher temperature. On moist litmus paper it shows an acid reaction. The substance does not change in color on drying at 105° C.

For combustion, the substance was mixed with fine copper oxide. After drying at 105° C. in vacuum over phosphorus pentoxide it was analyzed.

The needle-shaped crystals gave the following:

0.1534 gm. subst. gave 0.0901 gm. H_2O and 0.3329 gm. CO_2.
0.2944 gm. subst. gave 0.0702 gm. $Mg_2P_2O_7$.
0.1175 gm. subst. gave 6.75 c.c. nitrogen at 22° C. and 763 mm.
0.1688 gm. subst. lost 0.0154 gm. H_2O on drying.
0.1982 gm. subst. lost 0.0183 gm. H_2O on drying.
Found: C = 59.18; H = 6.57; P = 6.64; N = 6.56; H_2O = 9.12 and 9.23 per ct.

The plate-shaped crystals gave the following:

0.1786 gm. subst. gave 0.0993 gm. H_2O and 0.3909 gm. CO_2.
0.4972 gm. subst. gave 0.1089 gm. $Mg_2P_2O_7$.
0.1955 gm. subst. lost 0.0256 gm. H_2O on drying.
0.5997 gm. subst. lost 0.0783 gm. H_2O on drying.
Found: C = 59.60; H = 6.22; P = 6.10; H_2O = 13.09 and 13.05 per ct.

These compounds apparently do not represent any definite strychnine salt of inosite triphosphate but it would seem as if they were mixtures of the tri- and tetrastrychnine salts.

For tri-strychnine inosite triphosphate, $C_6H_{15}O_{15}P_3 (C_{21}H_{22}N_2O_2)_3$ = 1422.

Calculated: C = 58.23; H = 5.70; P = 6.54; N = 5.90 per ct.

For tetra-strychnine inosite triphosphate, $C_6H_{15}O_{15}P_3 (C_{21}H_{22}N_2O_2)_4$ = 1756.

Calculated: C = 61.50; H = 5.86; P = 5.30; N = 6.37 per ct.

PREPARATION OF THE BARIUM SALT FROM THE STRYCHNINE SALT.

The recrystallized strychnine salt, 27 grams, was dissolved in about 750 c.c. of hot water and the solution rendered alkaline with ammonia. After standing in ice water for some time the strychnine was filtered off. The filtrate was shaken with several portions of chloroform to remove the last traces of strychnine.

The solution, which contained the ammonium salt of the organic phosphoric acid, was precipitated by adding a solution of barium chloride in excess. After settling over night, the precipitate was

filtered and washed several times with water. It was then dissolved
in 1 per ct. hydrochloric acid, filtered, and precipitated with barium
hydroxide in excess. The precipitate was filtered and washed with
water until free from chlorides. It was again dissolved in 1 per ct.
hydrochloric acid, filtered, and the solution rendered neutral to
litmus with barium hydroxide. The precipitate was filtered and
washed free of chlorides with water and then in alcohol and ether
and dried in vacuum over sulphuric acid. The substance was a
pure white, amorphous powder. On moist litmus paper it showed
a very faint acid reaction. It was free from nitrogen.

For analysis it was dried at 100° in vacuum over phosphorus
pentoxide.

0.2323 gm. subst. gave 0.0370 gm. H_2O and 0.0758 gm. CO_2.
0.2085 gm. subst. gave 0.1668 gm. $BaSO_4$ and 0.0868 gm. $Mg_2P_2O_7$.
Found: C = 8.89; H = 1.78; P = 11.60; Ba = 47.07 per ct.

The substance was again dissolved in 1 per ct. hydrochloric acid
and the solution neutralized to litmus with barium hydroxide. It
was filtered, washed free of chlorides with water and then in alcohol
and ether and dried in vacuum over sulphuric acid.

It was analyzed after drying as before and the following results
obtained:

0.3168 gm. subst. gave 0.0481 gm. H_2O and 0.0972 gm. CO_2.
0.2366 gm. subst. gave 0.1968 gm. $BaSO_4$ and 0.0960 gm. $Mg_2P_2O_7$.
Found: C = 8.36; H = 1.69; P = 11.31; Ba = 48.94 per ct.

For the neutral barium salt of inosite triphosphate, $C_6H_9O_{15}P_3$
$Ba_3 = 826$.
Calculated: C = 8.71; H = 1.08; P = 11.25; Ba = 49.88 per ct.

In the two analyses reported above the carbon is somewhat low.
It must be noted, however, that these barium salts burned with
extreme difficulty. Traces of carbon remained after prolonged heating in a current of oxygen. The residues were mixed with fine copper
oxide and re-burned when one or two milligrams of carbon dioxide
were obtained, but we believe that the combustion even under these
conditions was incomplete.

PREPARATION OF THE ACID BARIUM SALT.

The above neutral barium salt was dissolved in the minimum
quantity of 1 per ct. hydrochloric acid, the solution was filtered, and
precipitated by adding about an equal volume of alcohol. The
resulting precipitate was filtered, washed free of chlorides with dilute
alcohol and then in absolute alcohol and ether and dried in vacuum
over sulphuric acid. The substance was then a pure white, amorphous powder which showed a strong acid reaction on moist litmus
paper.

It was analyzed after drying at 105° in vacuum over phosphorus pentoxide.

0.2588 gm. subst. gave 0.0538 gm. H_2O and 0.1046 gm. CO_2.
0.1691 gm. subst. gave 0.1027 gm. $BaSO_4$ and 0.0890 gm. $Mg_2P_2O_7$.
Found: C = 11.02; H = 2.32; P = 14.67; Ba = 35.74 per ct.

In the combustion of this substance a practically white ash was obtained.

This compound is evidently a complex acid salt of inosite triphosphate and agrees with the following formula:

$C_{18}H_{35}O_{45}P_9Ba_5 = 1937$.

(Calculated: C = 11.15; H = 1.80; P = 14.40; Ba = 35.46 per ct.) which may be graphically represented as follows:

$$C_6H_{12}O_{15}P_3 = Ba$$
$$> Ba$$
$$C_6H_{11}O_{15}P_3 = Ba$$
$$> Ba$$
$$C_6H_{12}O_{15}P_3 = Ba$$

That is, three molecules of mono-barium inosite triphosphate joined by two atoms of barium. Whether it is a compound as represented above or a mixture of various acid salts of inosite triphosphate can hardly be determined.

PREPARATION OF THE FREE ACID.

The barium salt from above was suspended in water and the barium precipitated with a slight excess of dilute sulphuric acid. The barium sulphate was filtered off and the filtrate precipitated with excess of copper acetate. The copper precipitate was filtered and washed with water until it gave no reaction with barium chloride. It was then suspended in water and decomposed with hydrogen sulphide. The copper sulphide was filtered off and the filtrate evaporated in vacuum at a temperature of 40°–45° to small bulk and then dried in vacuum over sulphuric acid. The substance was then obtained as a practically colorless syrup. After continued drying it forms a hard, sticky hygroscopic mass. It is extremely soluble in water and also readily, soluble in alcohol and in absolute alcohol. Much time was consumed in endeavoring to obtain it in crystalline form but without success. The syrupy substance was therefore analyzed after drying first for several days in vacuum over sulphuric acid at room temperature and finally at 100° C. in vacuum over phosphorus pentoxide. On drying at this temperature it turned quite dark in color.

0.1685 gm. subst. gave 0.0557 gm. H_2O and 0.1054 gm. CO_2.
0.1693 gm. subst. gave 0.1363 gm. $Mg_2P_2O_7$

Found: $C = 17.06$; $H = 3.69$; $P = 22.44$ per ct.
For inosite triphosphoric acid $C_6H_{15}O_{15}P_3 = 420$.
Calculated: $C = 17.14$; $H = 3.57$; $P = 22.14$ per ct.

In the above combustion a slight residue of unburned carbon remained enclosed in the fused metaphosphoric acid. It was mixed with some fine copper oxide and again burned when a few additional milligrams of carbon dioxide was obtained.

PROPERTIES OF INOSITE TRIPHOSPHORIC ACID.

The reactions of this acid differ in several particulars from phytic acid or inosite hexaphosphate.

The concentrated aqueous solution gives no precipitate with ammonium molybdate either in the cold or on warming. The cold aqueous solution of inosite hexaphosphate gives a white crystalline precipitate with ammonium molybdate.

The aqueous solution of the acid is not precipitated with silver nitrate. Inosite hexaphosphate gives a white amorphous precipitate with silver nitrate in excess. However, a solution of inosite triphosphate neutralized with ammonia gives a white amorphous precipitate with silver nitrate.

An aqueous solution of inosite triphosphoric acid when added to a solution of egg albumen causes only a slight turbidity — on longer standing a white precipitate separates slowly. Inosite hexaphosphate precipitates egg albumen immediately.

The acid is very soluble in water and readily soluble in alcohol and absolute alcohol — from the latter solution it is precipitated by ether in small oily drops.

The acid is not precipitated by barium or calcium chlorides but alcohol produces in these solutions white amorphous precipitates; these salts are likewise precipitated with ammonia.

CLEAVAGE OF INOSITE TRIPHOSPHORIC ACID INTO INOSITE AND PHOSPHORIC ACID.

One gram of the acid, dissolved in a little water, was heated with 10 c.c. of $\frac{5N}{4}$ sulphuric acid in a sealed tube for three hours to 150°–155°. The contents of the tube were then slightly yellowish brown in color. The sulphuric and phosphoric acids were precipitated with barium hydroxide and the inosite isolated in the usual way. Unfortunately a portion of the solution was lost but from what remained 0.15 gram inosite was obtained. After twice recrystallizing from dilute alcohol with addition of ether, 0.12 gram of inosite in the characteristic needle-shaped crystals was obtained. It gave the reaction of Scherer and melted at 222° C. (uncorrected). After mixing with some previously isolated and analyzed inosite the melting point did not change. The substance was therefore undoubtedly inosite and the analysis was omitted.

Although the barium salts described in this paper were amorphous and the inosite triphosphoric acid itself was a non-crystallizable syrup we believe that the substances were pure. The basis for this belief rests upon the fact that they had been prepared from the recrystallized strychnine salt which appeared perfectly homogeneous.

We have been unable to complete the investigation of the water-soluble portion of the barium salt mentioned on page 7. We wish, however, to record the following data although it is very incomplete.

After the aqueous solution had been precipitated with barium hydroxide, as already mentioned, the precipitate was filtered and washed in water. It was then dissolved in the minimum quantity of about 2 per ct. hydrochloric acid, filtered and precipitated with about an equal volume of alcohol. The voluminous amorphous precipitate was filtered and washed free of chlorides with dilute alcohol and then in absolute alcohol and ether and dried in vacuum over sulphuric acid. The dry substance weighed 38 grams.

It was again rubbed up in a mortar with about 200 c.c. of cold water, allowed to stand a short while and then filtered, washed in water, alcohol and ether and dried in vacuum over sulphuric acid. The water-insoluble portion weighed 10.3 grams.

The filtrate was concentrated in vacuum at a temperature of 40° C. to about 100 c.c. As the concentration proceeded a small quantity of a heavy white barium salt separated. It was not definitely crystalline and it only weighed 0.9 gram. The aqueous solution, about 100 c.c., contained therefore about 27 grams of the original substance. It was found impossible to obtain anything crystalline from this solution. However, on heating to boiling a heavy semi-crystalline or granular powder separated. This was filtered and washed in hot water, alcohol and ether and dried in the air. It weighed 1.7 grams.

The filtrate from above was precipitated by adding about half a volume of alcohol. The amorphous, white precipitate was filtered, washed and dried in vacuum over sulphuric acid.

A further quantity of substance was obtained by adding more alcohol to the filtrate which was likewise filtered, washed and dried.

These three fractions were analyzed after drying at 105° in vacuum over phosphorus pentoxide.

The semi-crystalline or granular substance gave the following:

$C = 10.14$; $H = 1.90$; $P = 13.18$; $Ba = 39.60$; $H_2O = 12.11$ per ct.

These results agree very closely with a dibarium salt of inosite triphosphate:

For $C_6H_{11}O_{15}P_3Ba_2$.

Calculated: $C = 10.41$; $H = 1.59$; $P = 13.45$; $Ba = 39.79$ per ct.

The first amorphous precipitate gave:

$C = 12.07$; $H = 2.30$; $P = 13.93$; $Ba = 33.70$ per ct.

The second amorphous precipitate gave:

C = 13.54; H = 2.74; P = 13.32; Ba = 32.38 per ct.

Lack of time has prevented the further examination of these substances but we hope to complete the investigation later.

From the results recorded it is evident that the organic phosphoric acids contained in the preparation used in this investigation consisted principally of inosite triphosphoric acid. The amorphous barium salts analyzed above may have contained some inosite diphosphoric acid or some other organic phosphoric acid. In addition to the above it must be noted that some inosite monophosphate described in an earlier paper had been isolated from this same wheat bran.

The author desires to express his appreciation and thanks to His Excellency Prof. E. Fischer for the interest shown in the work reported in this paper.

II. THE HYDROLYSIS OF PHYTIN BY THE ENZYME PHYTASE CONTAINED IN WHEAT BRAN.*

(Twelfth paper on phytin.)

INTRODUCTION.

It has been shown by the investigations of Suzuki, Yoshimura and Takaishi [1] that rice bran contains an enzyme which rapidly hydrolyzes phytin with formation of inosite and inorganic phosphoric acid. These authors concluded that wheat bran likewise contained a similar enzyme because the inorganic phosphorus increased in wheat bran extracts on standing.

Plimmer [2] examined a large number of extracts prepared from the intestines, liver, pancreas, castor beans, etc., as to their action on organic phosphorus compounds. While some of these showed a slight cleavage action on phytin, none could be compared in activity to an aqueous extract of wheat bran. The hydrolytic action of these extracts was determined by estimating from time to time the amount of inorganic phosphorus split off from phytin solutions of known concentration.

Since the above experiments clearly demonstrated that large quantities of inorganic phosphate were liberated from phytin by wheat bran extracts, it appeared of interest to determine what products, in addition to inosite and inorganic phosphoric acid, were formed under these conditions. For this purpose wheat bran extract was

---

*The work reported in this paper was carried out in the Ludwig Mond Biochemical Research Laboratory of the Institute of Physiology, University College, London.

[1] Bull. Coll. Agric. Tokyo, 7:503–512. 1907.

[2] Biochem. Journ. 7:43, 1913.

allowed to act upon a dilute solution of phytin at a temperature of 37°. Inorganic phosphoric acid was determined in the solution from time to time.

It was found that about two-thirds of the total phosphorus was split off during the first 16 days. Afterwards there was no appreciable change even on standing for about two years.

The solution had been prepared and the original determinations made by Dr. Plimmer. At his suggestion the writer undertook to examine the final reaction mixture for such products as had been formed.

These products were separated into two portions by precipitating the original solution with barium hydroxide. The precipitate contained inorganic barium phosphate, and also those barium salts of organic phosphoric acids that were insoluble in the dilute alkaline solution. The filtrate on the other hand was found to contain inosite monophosphoric acid and free inosite.

The inorganic phosphate and other impurities were removed from the crude barium hydroxide precipitate as will be described in the experimental part. The organic phosphoric acids which remained were obtained as amorphous barium salts. It was impossible to isolate any unchanged barium phytate. It is evident then that all of the phytin had been partially hydrolyzed.

The above amorphous substance appeared to consist mainly of barium inosite triphosphate, but probably mixed with some barium inosite diphosphate. Owing to the difficulty of separating these compounds their isolation was not attempted.

Among the soluble substances which had been formed we were able to isolate and identify inosite monophosphoric acid, a substance which we have previously isolated from wheat bran.[3] In addition to this, the solution also contained some free inosite which was isolated by means of its lead compound.

The action of this enzyme phytase upon phytin appears to proceed in several stages. Only a portion of the phytin is completely decomposed into inosite and phosphoric acid, but all of the phytin is partially hydrolyzed with formation of certain lower phosphoric acid esters of inosite, viz., inosite tri–, di– and monophosphoric acid and inorganic phosphoric acid. The formation of these intermediate products is only possible through the destruction or inhibition of the enzyme before the hydrolysis is complete. The reason for this inhibition is not clear but it may be due to the excess of phosphoric acid which is liberated.

It is interesting to note, and we call particular attention to the fact, that the organic phosphoric acids which remain as intermediate products of the action of the enzyme upon phytin, viz., inosite triphosphoric acid and inosite monophosphoric acid, are identical with

---

[3] *Journ. Biol. Chem.* 18:441, 1914; and N. Y. Agrl. Exp. Sta. Tech. Bull. 36, 1914.

the substances which we have isolated previously from wheat bran after it has been digested in 0.2 per ct. hydrochloric acid.

EXPERIMENTAL PART.

Commercial phytin, 100 grams, was dissolved in 500 c.c. of water and filtered from the insoluble matter which weighed 3.5 grams when dried at 100°. The pale yellow solution was treated with 38 grams of oxalic acid dissolved in about 250 c.c. of water. The calcium oxalate was filtered off, washed and dried. It weighed 48 grams. The solution was diluted to 6000 c.c. with water and was then found to contain 40 grams of phosphorus pentoxide. To it were added 800 c.c. of an aqueous extract of wheat bran which contained 2.2 grams of P_2O_5. The solution was kept under toluol at a temperature of 37°. No hydrolysis occurred in a week. This was evidently due to the strong acid reaction of the solution. It was nearly neutralized with ammonia and 735 c.c. of bran extract containing 1.53 grams of P_2O_5 were added. It was again kept at a temperature of 37° under toluol. In 9 days one-half of the total phosphorus was hydrolyzed; in 16 days two-thirds was hydrolyzed. In 35 days the amount of hydrolysis had not altered and after about two years it was again the same. The total and inorganic phosphorus was determined as described by Plimmer and Page.[4]

The dark-colored solution was filtered and barium hydroxide (Kahlbaum) added in slight excess. After standing over night the precipitate was filtered and washed in water. The filtrate and washings were evaporated on the water bath and the residue was examined as will be described later.

The barium precipitate was dissolved in about 2.5 per ct. hydrochloric acid, filtered, and precipitated by adding about an equal volume of alcohol. The precipitate was filtered and washed in dilute alcohol. The substance was again precipitated four times in the same way. It was then precipitated by barium hydroxide three times from about 2 per ct. hydrochloric acid and finally two times more with alcohol from the same strength hydrochloric acid. After the final filtering it was washed free of chlorides in dilute alcohol and then in alcohol and ether and dried in vacuum over sulphuric acid. The substance was then a snow-white, amorphous powder. It weighed 28.4 grams. It was free from chlorides and inorganic phosphate, and bases other than barium could not be detected.

The substance was then rubbed up in a mortar with 300 c.c. of cold water and allowed to stand with occasional shaking for a few hours. It was then filtered and washed in water, alcohol and ether, and dried in vacuum over sulphuric acid. The dry water-insoluble portion weighed 5.4 grams.

---

[4] *Biochem. Journ.* 7:162, 1913.

The filtrate from above was neutralized to litmus with barium hydroxide. The precipitate was filtered and washed in water, alcohol and ether and dried in vacuum over sulphuric acid. It weighed 23.6 grams.

These precipitates were analyzed after drying at 105° in vacuum over phosphorus pentoxide.

The first, water-insoluble portion, gave the following result:

Found : C=11.46 : H=1.93; P=11.59; Ba=39.94 per ct.

This substance is apparently largely composed of the di-barium inosite triphosphate; calculated for this, $C_6H_{11}O_{15}P_3Ba_2=690$.

C=10.43; H=1.59; P=13.47; Ba=39.71 per ct. It is, however, not pure but apparently contains some barium inosite diphosphate because the carbon is high and the phosphorus is low.

The water-soluble substance which was precipitated with barium hydroxide gave the following:

Found: C=9.63; H=1.63; P=10.91; Ba=47.41 per ct.

This substance also appears to consist largely of the neutral barium salt of inosite triphosphate. Calculated for the latter $C_6H_9O_{15}P_3Ba_3=826$. C=8.71; H=1.08; P=11.25; Ba=49.88 per ct. The carbon however is high and the phosphorus as well as barium are low which points to the presence of barium inosite di-phosphate.

In the hope of approximately separating these barium inosite tri- and diphosphates, the substance, 23.6 grams, was digested in dilute acetic acid for several hours with occasional shaking. It was then filtered and washed in water and the insoluble portion dried in vacuum over sulphuric acid. It weighed 10 grams.

The filtrate and washings containing the soluble portion of the substance was precipitated by adding lead acetate in excess. After standing over night the white, amorphous precipitate was filtered and washed in water. It was suspended in water and decomposed by hydrogen sulphide, filtered, and the excess of hydrogen sulphide boiled off. It was again precipitated in the same manner with lead acetate and decomposed with hydrogen sulphide. The solution still contained a considerable quantity of barium. The barium was therefore removed with a slight excess of dilute sulphuric acid. After filtering off the barium sulphate the solution was precipitated by adding copper acetate in excess. The copper precipitate was filtered, washed and suspended in water and decomposed with hydrogen sulphide. After removing the copper sulphide, the filtrate was evaporated in vacuum to small bulk and finally dried in vacuum over sulphuric acid. There remained a thick, nearly colorless syrup. It was readily soluble in alcohol. The addition of chloroform to this solution caused the substance to separate in small oily drops; the addition of ether produced a cloudiness, and on standing a flocculent, amorphous precipitate separated. These solutions could not be brought to crystallize. The acid preparation itself was kept for several weeks in the desiccator over sulphuric acid. It became a hard but sticky mass but showed absolutely no tendency to crystallize. Kept in this manner the color of the preparation gradually darkened.

Since the acid would not crystallize, the syrupy substance was analyzed after drying at 105° in vacuum ·over phosphorus pentoxide.

Found: C=18.58; H=3.82; P=20.38 per ct.
For inosite triphosphoric acid, $C_6H_{15}O_{15}P_3=420$
Calculated: C=17.14; H=3.57; P=22.14 per ct.
For inosite diphosphoric acid, $C_6H_{14}O_{12}P_2=340$
Calculated: C=21.17; H=4.11; P=18.23 per ct.

This acid preparation is evidently also a mixture of the inosite tri- and diphosphoric acids.

EXAMINATION OF THE FILTRATE AFTER THE WATER-INSOLUBLE BARIUM SALTS HAD BEEN PRECIPITATED.

The filtrate was evaporated, as mentioned on p. 15 and the residue taken up in hot water. It was decolorized with animal charcoal. The solution was neutral in reaction. It strongly reduced Fehling's solution on boiling — possibly due to sugars introduced with the bran extract. The solution was found to contain barium and also phosphorus in organic combination — evidently inosite monophosphate. The aqueous solution was precipitated by adding about an equal volume of alcohol and the white, amorphous precipitate filtered off — the filtrate being reserved for further examination.

Isolation of inosite monophosphate.— The above precipitate, which formed on the addition of alcohol, was dissolved in water, slightly acidified with acetic acid, and then precipitated with lead acetate in excess. After settling, this was filtered, washed in water and suspended in hot water and decomposed with hydrogen sulphide. It was then filtered and the filtrate boiled to expel excess of hydrogen sulphide. It was re-precipitated several times with lead acetate in the same manner until a white lead precipitate was obtained. This was finally decomposed with hydrogen sulphide, filtered, and evaporated to small bulk in vacuum and then dried in vacuum over sulphuric acid until a thick syrup remained. On scratching with a glass rod, this crystallized to a white solid mass. It was digested in alcohol and filtered, washed in alcohol and ether, and dried in the air. It weighed 1.6 grams. It had all the properties of inosite monophosphate. For further purification it was dissolved in a few c.c. of water and filtered. Alcohol was then added until the solution turned cloudy; it was heated until it cleared up and more alcohol was added until a faint permanent cloudiness remained. It was allowed to stand for about 48 hours at room temperature, when the substance had separated in massive, practically colorless crystals. After filtering, washing in alcohol and ether, and drying in the air,

1 gram substance was obtained. When heated in a capillary tube it began to soften at 188°–189° and melted under decomposition and effervescence at 190° C. (uncorrected). The appearance and properties of the substance corresponded exactly with those described for inosite monophosphate and the analysis was therefore omitted.

Isolation of inosite.— The filtrate, after precipitating the above barium salt of inosite monophosphate with alcohol, was evaporated on the water-bath until the alcohol was removed. It still contained barium, chlorides, etc. The barium was quantitatively precipitated with dilute sulphuric acid and the solution again concentrated on the water-bath.

The addition of lead acetate caused no precipitate. Basic lead acetate was then added so long as any precipitate formed. This precipitate was filtered off and discarded. The solution was then heated to boiling and more basic lead acetate added and the solution was finally made strongly alkaline with ammonia and allowed to stand over night. This precipitate was filtered and washed in water and then decomposed in aqueous suspension with hydrogen sulphide. The filtrate was concentrated on the water-bath and the inosite brought to crystallization by the addition of alcohol. After recrystallizing several times 0.5 gram of pure inosite was obtained in the characteristic needle-shaped crystals. It gave the reaction of Scherer and melted at 218° C. (uncorrected).

III. THE HYDROLYSIS OF THE ORGANIC PHOSPHORUS COMPOUND OF WHEAT BRAN BY THE ENZYME PHYTASE.*

(Thirteenth paper on phytin.)

INTRODUCTION.

In several previous papers[1] on the subject of the organic phosphoric acid compounds of wheat bran we have mentioned that when wheat bran is digested in 0.2 per ct. hydrochloric acid, the resulting extract always contains relatively large quantities of inorganic phosphate. Up to the present we have had no data concerning the origin of this inorganic phosphate and it remained to determine whether it was originally present in the bran or if it had been formed during the digestion by hydrolysis from the organic phosphorus compound.

*The work reported in this paper was carried out in the Ludwig Mond Biochemical Research Laboratory of the Institute of Physiology, University College, London.

[1] *Journ. Biol. Chem.* 12:447, 1912; 18:425, 1914; and N. Y. Agrl. Expt. Sta. Tech. Bull. 22 and 36.

Hart and Andrews,[2] who examined wheat bran as well as a large number of other feeding materials, came to the conclusion that the phosphorus was present almost entirely in organic combination. Some criticism; however, has been offered of their method of estimating inorganic phosphorus and it has been suggested that the time, 15 minutes, which they allowed for digestion was not sufficient for complete extraction.[3]

From the work of Suzuki, Yoshimura and Takaishi[4] and of Plimmer[5] it is evident that wheat bran contains an enzyme which rapidly hydrolyzes a portion of the organic phosphorus into inorganic phosphoric acid. The quantitative determinations of the activity of this enzyme on phytin or inosite hexaphosphate reported in the preceding paper shows that some two-thirds of the total phosphorus of the phytin was eliminated as inorganic phosphoric acid in about 16 days. Since a small quantity of wheat bran extract is capable of hydrolyzing phytin to such an extent, it is evident that the organic phosphoric acid compound originally present in the bran would also be hydrolyzed on digestion under the same conditions.

The determinations herein reported were undertaken in the hope of throwing some light upon this question. It was also hoped to determine the cause of the large percentage of inorganic phosphoric acid in extracts obtained on digesting wheat bran in 0.2 per ct. hydrochloric acid.

The results show that the enzyme phytase contained in the wheat bran very rapidly hydrolyzes the organic phosphorus compound of the bran into inorganic phosphoric acid under certain conditions. We have endeavoured to determine under what conditions the maximum activity of the enzyme manifests itself and also the conditions under which its action is either inhibited or destroyed.

When wheat bran is digested in distilled water the hydrolysis begins at once and proceeds with considerable rapidity, and at the end of 24 hours nearly 90 per ct. of the total water-soluble phosphorus is inorganic. The organic and inorganic phosphates of the bran are only partially soluble in the water, and when the extracts, after digesting the bran for 24 hours in water, are acidified with hydrochloric acid, then only some 60 per ct. of the total dissolved phosphorus compounds are inorganic. (See Table II). This result is not perceptibly altered by digesting the bran in water for 48 hours and then acidifying. (See Table III.)

When the bran is digested in 0.2 per ct. hydrochloric acid the rate of hydrolysis is greater than with pure water. The actual amounts of both organic and inorganic phosphorus are greater in the dilute acid than in the aqueous extracts. After 3 hours about one-third of the

[2] N. Y. Agrl. Expt. Sta. Bull. 238, 1903.
[3] Ohio Agrl. Expt. Sta. Bull. 215.
[4] *Bull. Coll. Agric. Tokyo*, 1907, 7:503–512, 1907.
[5] *Biochem. Journ.* 7:43, 1913.

total phosphorus is inorganic and after 24 hours about three-quarters of the total soluble phosphorus has been hydrolyzed to inorganic phosphoric acid, as is shown in Table V.

With 0.2 per ct. acetic acid the hydrolysis is more rapid and more complete than with 0.2 per ct. hydrochloric acid, 85 per ct. in 24 hours (Table VII) and the most complete hydrolysis was obtained with 0.1 per ct. hydrochloric acid, namely, 92 per ct. (Table VIII.)

That the hydrolysis of the organic phosphorus compound is due to enzyme action and not to the acid employed is strikingly illustrated by the fact that with increasing strengths of the hydrochloric acid the amount of inorganic phosphorus rapidly and steadily diminishes. But when 0.5, 1.0 or 2.0 per ct. hydrochloric acid is used, the inorganic phosphorus is practically constant. (Table V.)

The enzyme is remarkably sensitive to certain concentrations of hydrochloric acid. The hydrolyzing action is stimulated by the presence of 0.1 or 0.2 per ct. hydrochloric acid much above that of pure water, but 0.3 per ct. almost inhibits the enzyme activity. With the 0.1 per ct. acid 92 per ct. and with the 0.2 per ct. acid 76 per ct. of the total phosphorus is inorganic, but with 0.3 per ct. acid only about 20 per ct. of the total phosphorus is inorganic. (See Tables V and VIII.)

Since such a slight difference in strength of the hydrochloric acid caused such a great difference in the amount of the inorganic phosphorus, it appeared of interest to determine whether the acid merely inhibited the hydrolysis or if the enzyme was destroyed by it. Some of the bran was digested in 0.5 per ct. hydrochloric acid for $1\frac{1}{2}$ hours and the acid then nearly neutralized with ammonia and the whole allowed to digest for 24 hours. In this case there was no increase in the inorganic phosphorus, showing that hydrochloric acid of this strength completely destroys the enzyme. (Table X.)

The enzyme is also destroyed by digesting the bran in 0.25 per ct. ammonia, as well as by pouring boiling water or boiling 0.2 per ct. hydrochloric acid over the bran. (Tables X and IX.)

Concerning the amount of inorganic phosphorus in wheat bran, it is evident, judging by the results obtained with 0.5, 1.0 and 2.0 per ct. hydrochloric acid as well as on treating bran with boiling water or boiling 0.2 per ct. hydrochloric acid with 0.25 per ct. ammonia, that ordinary bran contains about 0.1 per ct. of phosphorus as inorganic, which represents about 11 per ct. of the total soluble phosphorus. It seems quite certain that under these conditions of extraction no hydrolysis of the organic phosphorus occurs, and we believe that these figures represent the amount of inorganic phosphorus normally present in wheat bran.

These results are not at variance with the findings of Hart and Andrews.[6] After digesting bran in 1.0 per ct. hydrochloric acid for 40 hours, they found 0.087 per ct. of inorganic phosphorus in the extract. This is practically the same as our results, 0.081 per ct.

[6] Loc. cit.

inorganic phosphorus, after digesting bran for 5 hours in 1.0 per ct. hydrochloric acid. The same authors found only 0.036 per ct. inorganic phosphorus in wheat bran after digesting in 0.2 per ct. hydrochloric acid for 15 minutes. The lower percentage is no doubt due to incomplete extraction and we believe that the higher figures represent the actual amount present.

EXPERIMENTAL PART.

Total and inorganic phosphorus were determined in extracts prepared from wheat bran. The extracts were prepared by digesting wheat bran in water or dilute acid for varying lengths of time; the particulars and details being given at the beginning of each table. Total phosphorus was determined after decomposing by the Neumann method. Inorganic phosphorus was determined as follows: The extract was diluted with some 50 c.c. of water, 15 grams of ammonium nitrate were added and then warmed to 65° on the water bath. It was then strongly acidified with nitric acid and the phosphorus precipitated with ammonium molybdate and the solution kept at the above temperature for half an hour. Under these conditions there is no danger of cleavage of the organic phosphorus — at least weighable quantities of phosphorus are not precipitated during this time from preparations which are free from inorganic phosphate. Plimmer[7] has also shown that it is necessary to heat phytin solutions with $\frac{N}{1}$ or $\frac{2N}{1}$ nitric acid for several hours to a temperature of 75° or higher before appreciable quantities of inorganic phosphorus are split off.

The phosphomolybdate precipitate was then filtered off and the phosphorus determined as magnesium pyrophosphate in the usual way.

The digestions were made throughout these experiments at room temperature, about 16° C.

Ten grams of the bran were digested in 100 c.c. of water for the time mentioned and then filtered and 20 c.c. used for each determination. The same quantities were used throughout.

TABLE I.— CHANGES IN CONDITION OF PHOSPHORUS IN WHEAT BRAN BY WATER DIGESTION.

Time.	Total phosphorus in extract.	Inorganic phosphorus in extract.	Part of total phosphorus in inorganic form.
	Per ct.	Per ct.	Per ct.
½ hour..................	· 0.362	0.165	45.77
1 "	0.420	0.239	56.95
4 "	0.508	0.383	75.34
20 "	0.655	0.576	88.00
24 "	0.747	0.660	88.43

[7] *Biochem. Journ.* 7:72, 1913.

It will be noticed that the hydrolysis is greatest during the first four hours. The percentage of phosphorus represents only that amount which was soluble in the water.

After digesting in water for 24 hours and then acidifying with hydrochloric acid and shaking for half an hour the following results were obtained:

TABLE II.— CHANGES IN CONDITION OF PHOSPHORUS IN WHEAT BRAN AFTER WATER DIGESTION AND ACIDIFYING.

Time.	Total phosphorus in extract.	Inorganic phosphorus in extract.	Part of total phosphorus in inorganic form.
	Per ct.	Per ct.	Per ct.
24 hours.................	1.06	0.649	60.85

After digesting the bran for 48 hours in water and then adding 100 c.c. of 2 per ct. hydrochloric acid and allowing to stand for another 24 hours, the following results were obtained:

TABLE III.— CHANGES IN CONDITION OF PHOSPHORUS IN WHEAT BRAN AFTER WATER DIGESTION, ACIDIFYING AND STANDING.

Time.	Total phosphorus in extract.	Inorganic phosphorus in extract.	Part of total phosphorus in inorganic form.
Hrs.	Per ct.	Per ct.	Per ct.
48 in water, 24 in weak acid..	1.27	0.803	62.79
48 in water, 24 in weak acid..	1.28	0.796	62.13

The addition of toluol to the water appears to diminish the amount of phosphorus dissolved without affecting the degree of hydrolysis.

TABLE IV.— EFFECT OF TOLUOL ON HYDROLYSIS OF PHOSPHORUS COMPOUNDS IN WHEAT BRAN.

Time: with toluol.	Total phosphorus in extract.	Inorganic phosphorus in extract.	Part of total phosphorus in inorganic form.
Hrs.	Per ct.	Per ct.	Per ct.
24..................	0.518	0.462	89.25

Using hydrochloric acid as the extracting medium, the strength varying from 0.2 to 2.0 per ct., the following results were obtained:

10 grams of bran were digested in 100 c.c. of the acid for the length of time mentioned in the table. It was then filtered and 20 c.c. of the filtrate used for each determination.

TABLE V.— EFFECT OF CONCENTRATION OF ACID ON HYDROLYSIS OF PHOSPHORIC ACID COMPOUND IN WHEAT BRAN.

Solvent.	Time.	Total phosphorus in extract.	Inorganic phosphorus in extract.	Part of total phosphorus in inorganic form
HCl, per ct.	Hrs.	Per ct.	Per ct.	Per ct.
0.2..................	3	1.130	0.415	36.52
0.2..................	24	1.200	0.925	76.87
0.3..................	24	0.999	0.203	20.22
0.4..................	24	0.939	0.146	15.61
0.5..................	20	0.922	0.124	13.44
1.0..................	5	0.894	0.081	9.13
2.0..................	30	1.080	0.117	10.83

The presence of toluol did not materially alter these results. Hydrochloric acid, 0.5 per ct. after 20 hours, with toluol gave the following :

TABLE VI.— EFFECT OF TOLUOL ON HYDROLYSIS OF PHOSPHORUS COMPOUNDS IN WHEAT BRAN.

Time: with HCl and toluol.	Total phosphorus in extract.	Inorganic phosphorus in extract.	Part of total phosphorus in inorganic form.
Hrs.	Per ct.	Per ct.	Per ct.
20......................	0.890	0.117	13.14

When the bran is digested in dilute acetic acid the hydrolysis is greater than with the same strength hydrochloric acid, as shown by the following results:—

TABLE VII.— EFFECT OF ACETIC ACID ON HYDROLYSIS OF PHOSPHORIC ACID COMPOUND IN WHEAT BRAN.

Solvent.	Time.	Total phosphorus in extract.	Inorganic phosphorus in extract.	Part of total phosphorus in inorganic form.
Acetic acid, per ct.	Hrs.	Per ct.	Per ct.	Per ct.
0.2.................	3	1.21	0.695	57.03
0.2.................	24	1.23	1.050	85.54

The maximum activity of the enzyme, as shown by the greatest hydrolysis, was obtained by digesting bran for 48 hours in water and then adding 0.1 per ct. hydrochloric acid and allowing to stand for another 24 hours. Ten grams of bran were digested in 100 c.c. of water for 48 hours and then 100 c.c. of 0.2 per ct. hydrochloric acid added. After standing for 24 hours more it was filtered and 20 c.c. of the filtrate used for each determination.

TABLE VIII.— MAXIMUM HYDROLYSIS OF PHOSPHORIC ACID COMPOUND IN WHEAT BRAN.

Time.	Total phosphorus in extract.	Inorganic phosphorus in extract.	Part of total phosphorus in inorganic form.
Hrs.	Per ct.	Per ct.	Per ct.
72.	1.24	1.14	92.37

DESTRUCTION OF THE ENZYME "PHYTASE" BY HEAT.

As is shown by the following experiments the hydrolytic action of the enzyme is completely destroyed when bran is exposed to the action of boiling water or boiling dilute hydrochloric acid for a short time. Ten grams of bran were placed in an Erlenmeyer flask and 100 c.c. of boiling water poured over it. It was then heated for a few minutes until the water boiled. It was allowed to digest for 24 hours at room temperature. Another lot of bran was treated in the same way but using 0.2 per ct. hydrochloric acid instead of water. The resulting extracts could not be filtered. The whole was therefore diluted with 100 c.c. of 2 per ct. hydrochloric acid and allowed to stand a few minutes to settle. Lots of 20 c.c. of the liquid were then taken for each determination.

TABLE IX.— EFFECT OF BOILING WATER OR BOILING ACID ON HYDROLYSIS OF PHOSPHORIC ACID COMPOUND IN WHEAT BRAN.

Solvent.	Time.	Total phosphorus in extract.	Inorganic phosphorus in extract.	Part of total phosphorus in inorganic form.
	Hrs.	Per ct.	Per ct.	Per ct.
Boiling water	24	0.886	0.108	12.66
Boiling 0.2% HCl	24	1.140	0.105	9.22

The hydrolytic action of the enzyme is likewise destroyed by exposing bran for a short time to 0.5 per ct. hydrochloric acid or by digesting bran in dilute ammonia. Ten grams of bran were digested in 100 c.c. of 0.5 per ct. hydrochloric acid for 1½ hours. The acid was then nearly neutralized with ammonia, leaving the solution faintly acid, and then standing for 24 hours. Ten grams of bran were digested in 100 c.c. of 0.25 per ct. ammonia for 24 hours. It was then acidified with dilute hydrochloric acid.

TABLE X.—EFFECT OF STRONG ACID AND OF AMMONIA ON HYDROLYSIS OF PHOSPHORIC ACID COMPOUND IN WHEAT BRAN.

Solvent.	Time.	Total phosphorus in extract.	Inorganic phosphorus in extract.	Part of total phosphorus in inorganic form.
	Hrs.	Per ct.	Per ct.	Per ct.
0.5% HCl............	24	0.661	0.115	17.46
0.25% NH₃..........	24	0.922	0.087	9.51

The results reported in this and the preceding paper will naturally change the interpretation of our previous investigations concerning the nature of the organic phosphorus compound of wheat bran. We have heretofore been under the impression that the substances which we have isolated from wheat bran, after digesting in 0.2 per ct. hydrochloric acid, represented the compounds originally present in the bran. That this supposition is erroneous is evident, since the action of a bran extract on phytin gives rise to identical compounds.

A consideration of the results reported in Table V likewise shows that there is a wide difference between the action of 0.2 and 1.0 per ct. hydrochloric acid on bran. In the first case the extract contains 0.925 per ct. inorganic phosphorus and in the second only 0.081 per ct.

These facts lead to the conclusion that the products which we have previously isolated from wheat bran represent only such intermediate compounds as have been formed by the action of the enzyme phytase upon the original phytin of the bran.

IV. CONCERNING PHYTIN IN WHEAT BRAN.*

(Fourteenth paper on phytin.)

INTRODUCTION.

It has been shown in earlier reports[1] that the organic phosphorus compounds of wheat bran which have been isolated after digesting

---

*The work reported in this paper was carried out in the Ludwig Mond Biochemical Research Laboratory of the Institute of Physiology, University College, London.

[1] Journ. Biol. Chem. 12:447, 1912; also 18:425, 1914; and N. Y. Agrl. Expt. Sta. Tech. Bull. 22 and 36.

the bran in 0.2 per ct. hydrochloric acid differ in composition and properties from phytin or inosite hexaphosphate. It has been found that under these conditions several organic phosphoric acids are obtained. Of these we have isolated and identified two, viz., inosite monophosphoric acid[2] and inosite triphosphoric acid.[3]

Patten and Hart,[4] who first investigated the phosphorus compounds of wheat bran, came to the conclusion that the organic phosphoric acid in bran was identical with phytic acid or the "anhydro-oxymethylene diphosphoric acid" of Posternak. These authors isolated the acid preparation which they analyzed after digesting the bran in 0.2 per ct. hydrochloric acid. We have shown conclusively, however, that when wheat bran is digested in this strength hydrochloric acid, the organic phosphorus compounds finally isolated are entirely different in composition from phytin or inosite hexaphosphate.

By the same method as above, i.e., after digesting in 0.2 per ct. hydrochloric acid, we have isolated inosite hexaphosphate from corn,[5] cottonseed meal [5] and oats.[5] All of these preparations were found to be identical with the inosite hexaphosphate prepared from commercial phytin.[6]

It would appear, then, as if wheat bran is the only one of all the various plants and seeds examined which does not contain inosite hexaphosphate. Instead, certain lower inosite phosphoric acids appear to be present. It seems difficult to explain why wheat bran should contain different inosite phosphoric acids from other plants.

The work of Suzuki, Yoshimura and Takaishi[7] and of Plimmer[8] on the presence of an enzyme in wheat bran which rapidly hydrolyzes phytin with formation of inorganic phosphoric acid, particularly in connection with the two preceding investigations, offered a key to the solution of this problem. The fact that the action of a wheat bran extract on commercial phytin produced products identical with those which we have previously isolated from wheat bran itself, viz., principally inosite triphosphoric acid and inosite monophosphoric acid, led to the thought that the organic phosphorus compound originally present in bran was probably hydrolyzed during the extraction with 0.2 per ct. hydrochloric acid with formation of inorganic phosphoric acid and lower inosite phosphoric acids. This opinion was fully confirmed by the results reported in the preceding paper. It is shown there that inorganic phosphoric acid is liberated rapidly when wheat bran is digested in water, in 0.1 and in 0.2 per ct. hydrochloric acid, as well as in 0.2 per ct. acetic acid. This rapid forma-

[2] Journ. Biol. Chem. 18:441, 1914; and N. Y. Agrl. Expt. Sta. Tech. Bull. 36.
[3] See a preceding article.
[4] Amer. Chem. Journ. 31:566, 1904.
[5] Journ. Biol. Chem. 17:141, 151, 165, 1914; and N. Y. Agrl. Expt. Sta. Tech. Bull. 32.
[6] Ibid. 17:171, 1914; and N. Y. Agrl. Expt. Sta. Tech. Bull. 32.
[7] Bull. Coll. Agric. Tokyo. 7:503–512, 1907.
[8] Biochem. Journ. 7:43, 1913.

tion of inorganic phosphoric acid from the organic phosphorus compound would naturally preclude the possibility of isolating the original substance from extracts prepared by digesting the material in the above solutions. The only products possible of isolation would be such intermediate substances as had escaped the hydrolytic action of the enzyme. The previous determinations concerning the activity of the enzyme showed, however, that the use of hydrochloric acid stronger than 0.2 per ct. materially reduced the amount of inorganic phosphate in the extract and that the minimum quantity was present when the bran was extracted with 1 per ct. hydrochloric acid.

Wheat bran was therefore extracted for 5 hours with 1 per ct. hydrochloric acid and the organic phosphoric acid in the extract isolated as a barium salt (see Experimental Part). The substance then obtained had entirely different properties and composition from those obtained when wheat bran is digested in 0.2 per ct. hydrochloric acid. After careful purification this salt crystallized in the same form and under the same conditions as the corresponding barium salts of inosite hexaphosphate. So far as crystal form, properties and composition are concerned, there appears to be no difference between the substances isolated from wheat bran and the barium salts of phytic acid or inosite hexaphosphate obtained from other sources. We conclude therefore that all of these materials, viz., wheat bran, corn, oats, cottonseed meal and commercial phytin, contain the same organic phosphoric acid, namely, phytic acid or inosite hexaphosphoric acid, $C_6H_{18}O_{24}P_6$.

This confirms the conclusions of Patten and Hart[9] as to the nature of the organic phosphorus compound of wheat bran. It is somewhat difficult to understand how these authors came to this conclusion since they state that they had extracted the bran with 0.2 per ct. hydrochloric acid, and under these conditions we have found that phytic acid or inosite hexaphosphoric acid is not obtained but principally inosite triphosphoric acid and some inosite monophosphoric acid.

In view of the results reported in this as well as in the two preceding papers, it is evident that the compounds which we have previously isolated from wheat bran, viz., inosite triphosphoric acid and inosite monophosphoric acid, do not represent the organic phosphoric acids originally present in the bran but that they are intermediate products formed from inosite hexaphosphoric acid by the enzyme, "phytase," during the extraction of the bran with the dilute acid.

EXPERIMENTAL PART.

The bran, 700 grams, was digested in 5 litres of 1 per ct. hydrochloric acid for 5 hours. It was then strained through cheesecloth and the liquid filtered. Barium hydroxide (Kahlbaum) was added to the filtrate until the reaction was alkaline. The precipitate was

[9] Loc. cit.

filtered and washed with water and then dissolved in about 3 per ct. hydrochloric acid. The opalescent solution was filtered through charcoal and the filtrate precipitated by adding about an equal volume of alcohol. After standing over night the precipitate was filtered, washed in dilute alcohol and again dissolved in 3 per ct. hydrochloric acid and filtered through charcoal. A dilute solution of barium hydroxide was gradually added to the filtrate until a precipitate began to form. After standing over night the substance had separated out in semicrystalline form. It was filtered and washed in water, dissolved in 3 per ct. hydrochloric acid and precipitated by alcohol. After filtering and washing with dilute alcohol it was again dissolved in the dilute hydrochloric acid and precipitated by barium hydroxide.

The precipitate was still dark colored and it contained some impurities, not completely soluble in the dilute hydrochloric acid, apparently of colloidal nature, which could not be removed by filtration. In order to eliminate these impurities the barium precipitate was suspended in water and the barium removed with a slight excess of dilute sulphuric acid. The barium sulphate was filtered off and the filtrate precipitated with excess of copper acetate. The copper precipitate was filtered and washed free of sulphates with water. It was then suspended in water and decomposed with hydrogen sulphide. After filtering off the copper sulphide a clear and colorless solution of the free acid was obtained. By these various operations the oxalic acid had also been removed, as, after nearly neutralizing with barium hydroxide and adding barium chloride, no precipitate or turbidity occurred. The solution was precipitated with barium hydroxide, filtered and washed in water. The substance was again twice precipitated with barium hydroxide from 3 per ct. hydrochloric acid and finally twice precipitated with alcohol from the same strength hydrochloric acid. After finally filtering, the substance was washed free of chlorides with dilute alcohol, alcohol and ether, and dried in vacuum over sulphuric acid. It was a snow-white semi-crystalline powder which weighed 11 grams. It was free from chlorides and inorganic phosphate and did not contain any bases except barium.

For analysis it was recrystallized as follows: 2 grams of the substance were dissolved in the minimum quantity of 2 per ct. hydrochloric acid and the free acid nearly neutralized with barium hydroxide and the solution filtered. The clear filtrate was heated to boiling when the substance separated as a heavy crystalline powder. This was filtered and washed in boiling hot water and finally in alcohol and ether and dried in the air. Yield 1.5 grams. It was recrystallized a second time in the same manner except that 20 c.c. of $\frac{N}{1}$ barium chloride were added to the solution before boiling. After filtering, washing and drying, as before, about 1.3 grams substance was obtained. It consisted of fine, microscopic, needle-shaped crystals. It was free from chlorides and the nitric acid solution gave no precipitate with ammonium molybdate, showing that inorganic phosphate was absent.

It was analysed after drying at 105° in vacuum over phosphorus pentoxide.

0.3771 gm. subst. lost 0.0385 gm. H_2O
0.1917 gm. subst. lost 0.0196 gm. H_2O
0.3386 gm. subst. gave 0.0361 gm. H_2O and 0.0794 gm. CO_2
0.2571 gm. subst. gave 0.0269 gm. H_2O and 0.0617 gm. CO_2
0.1715 gm. subst. gave 0.1218 gm. $BaSO_4$ and 0.1026 gm. $Mg_2P_2O_7$
Found: C=6.39; H=1.19; P=16.67; Ba=41.79 per ct.
        C=6.54; H=1.17
        H_2O 10.20 and 10.22 per ct.

For heptabarium inosite hexaphosphate $(C_6H_{11}O_{24}P_6)_2Ba_7$=2267
Calculated: C=6.35; H=0.97; P=16.40; Ba=42.39 per ct.
For 14 H_2O calculated, 10.00 per ct.

PREPARATION OF THE CRYSTALLINE TRIBARIUM INOSITE HEXAPHOSPHATE.

This was prepared by dissolving 5 grams of the original substance in 2 per ct. hydrochloric acid, nearly neutralizing with dilute barium hydroxide, filtering and adding alcohol gradually until the solution turned cloudy. It was then allowed to stand for two days at room temperature. The substance separated slowly in the form of globular masses or rosettes of microscopic needles. The crystal form was identical with that previously described for the tribarium inosite hexaphosphate. The substance was filtered and washed in dilute alcohol, alcohol and ether, and dried in the air. It was recrystallized a second time by dissolving in the minimum quantity of 2 per ct. hydrochloric acid, filtering, and adding alcohol until a slight permanent cloudiness remained. After standing for two days the substance had separated in the same form as before. It was filtered, washed free of chlorides with dilute alcohol, and then in alcohol and ether and dried in the air. It was obtained as a snow-white crystalline powder. It was free from chlorides and inorganic phosphates.

For analysis it was dried at 105° in vacuum over phosphorus pentoxide.

0.2922 gm. subst. lost 0.0334 gm. H_2O
0.2588 gm. subst. gave 0.0291 gm. H_2O and 0.0648 gm. CO_2
0.1421 gm. subst. gave 0.0944 gm. $BaSO_4$ and 0.0867 gm. $Mg_2P_2O_7$
Found: C=6.82; H=1.25; P=17.00; Ba=39.09 per ct.
        H_2O=11.43 per ct.

For tribarium inosite hexaphosphate, $C_6H_{12}O_{24}P_6Ba_3$=1066
Calculated: C=6.75; H=1.12; P=17.44· Ba=38.65 per ct.
For 8 H_2O calculated 11.90 per ct.

Most authors working with organic phosphorus compounds report much difficulty in obtaining proper values for the carbon. This difficulty is particularly great in burning compounds such as those reported above where the percentage of carbon is so low. Under ordinary conditions it is impossible to obtain a complete combustion — the ash is usually more or less dark colored. Some authors recommend mixing the substance intimately with fine copper oxide. This procedure is very serviceable when burning salts of these organic phosphoric acids with organic bases — like the strychnine salts which we have previously reported, but with barium salts we have not found copper oxide to be of much use. In the analyses reported above we have used the following method, for the suggestion of which we are indebted to His Excellency Prof. E. Fischer of Berlin.

The substance is first burned in the usual way in a current of oxygen — the combustion lasting about an hour. The calcium chloride tube and the potash bulb are then weighed. The increase in weight of the calcium chloride tube is taken as the correct weight of the water. The residue in the boat, which is dark colored from particles of unburned carbon, is powdered in an agate mortar with some recently fused potassium bichromate and again placed in the boat, the mortar being rinsed out with some more powdered bichromate. The whole is again burned in the usual way. The potassium bichromate fuses and oxidizes all the carbon in the residue. The increase in weight in the potash bulb is added to the first, giving the total carbon dioxide.

Since the barium salts described above agree in crystal form and composition with salts of inosite hexaphosphoric acid or phytic acid, we believe there can be no doubt that wheat bran contains the same phytin as other plants. We would have recognized this relationship sooner if we had made a series of inorganic phosphorus determinations in wheat bran extracts prepared with water or dilute acids, such as is reported in a preceding paper. We believed, however, that since phytic acid could be isolated from cottonseed meal, oats and corn, after digesting in 0.2 per ct. hydrochloric acid, that the same procedure should also suffice in the case of wheat bran. Moreover, we were following the method of isolation recommended by previous investigators on this subject. The fact that inosite hexaphosphoric acid is obtained from 0.2 per ct. hydrochloric acid extracts of corn, oats, and cottonseed meal, only proves that these materials, with the possible exception of oats, do not contain any enzyme of the nature of the phytase contained in wheat bran. In the case of the phytin preparations isolated from oats[10] we found a considerable amount of water-soluble barium salt of practically the same composition as the barium salts prepared from wheat bran

[10] Journ. Biol. Chem. 17:163, 1914; N. Y. Agrl. Expt. Sta. Tech. Bull. 32, p. 21.

after extracting with 0.2 per ct. hydrochloric acid. We hope to investigate further this matter concerning the distribution of the enzyme phytase in various plants. We also intend to study more closely certain phases of the action of this enzyme, particularly whether it is capable of inducing synthetic reactions. It appears not improbable that an equilibrium exists in the reaction

$$C_6H_6O_6 [PO(OH)_2]_6 + 6H_2O \rightleftharpoons C_6H_6(OH)_6 + 6H_3PO_4$$

particularly at the stage when one-half of the phosphorus has been split off. The fact that the bulk of the substance ordinarily isolated from wheat bran, after digesting in 0.2 per ct. hydrochloric acid, is inosite triphosphate, supports this view.

The author acknowledges with pleasure his indebtedness to Dr. R. H. A. Plimmer for many suggestions which have been of great assistance in carrying out the work reported in this and the two preceding papers.

TECHNICAL BULLETIN No. 41. APRIL, 1915.

New York Agricultural Experiment Station.

GENEVA, N. Y.

FIBRIN.
ALFRED W. BOSWORTH

PUBLISHED BY THE DEPARTMENT OF AGRICULTURE.

FIBRIN.*

ALFRED W. BOSWORTH.

SUMMARY.

1. Fibrin can combine with both bases and acids to form definite compounds.

2. Fibrin combines with four equivalents of base to form a compound which is neutral to phenolphthalein.

3. Fibrin combines with bases to form a series of three acid salts which contain one, two, and three equivalents of base respectively.

4. All the combinations of fibrin with sodium, potassium, and ammonium are soluble.

5. The calcium fibrinates containing three and four equivalents of calcium are soluble, the calcium fibrinates containing one and two equivalents of calcium being insoluble.

6. Fibrin combined with one equivalent of acid is insoluble, and combined with more than one equivalent of acid is soluble.

7. Pure fibrin, unlike casein, is not strong enough as an acid to decompose calcium carbonate.

8. The molecular weight of fibrin is about 6666.

9. Carbon dioxide precipitates fibrin from a solution of calcium fibrinate, but not from a solution of sodium, potassium, or ammonium fibrinate.

INTRODUCTION.

Certain observations made while working with blood seemed to indicate that fibrin might possess some chemical properties quite similar to those of casein, which have been reported by Van Slyke and Bosworth.[1] Upon investigation it has been found that fibrin resembles casein in many of its reactions towards bases and acids, and it is thought by the author that these reactions may play an important rôle in the phenomenon of the clotting of blood. As a continuation of the work beyond the point reported in this bulletin seems to be out of the scope of this institution, it has seemed best to publish the results obtained in order that they might be available to others interested in the question of the clotting of blood. ▓

* Published in the *Journal of Biological Chemistry*, 20:91, 1915, as a contribution from the Boston Floating Hospital, Boston, and the Chemical Laboratory of the New York Agricultural Experiment Station, Geneva.
[1] *Jour. Biol. Chem.* 14:203–236, 1913, and Technical Bull. 26 of this Station.

EXPERIMENTAL.

METHOD OF PREPARING ASH-FREE FIBRIN.

Fresh ox blood was collected in a large bottle and carried immediately to the laboratory where it was transferred to precipitating jars and allowed to coagulate. The clots were removed, broken into small pieces, and washed in running water to remove the serum and blood corpuscles. The washed masses of fibrin were passed through a meat chopper, placed in an 8-liter bottle, a little toluene was added, and the bottle filled with a 0.2 per ct. solution of sodic hydrate. This solution caused the fibrin to swell, and after about 36 hours the whole contents of the bottle resembled a thin jelly. This jelly was broken up, one-half transferred to another 8-liter bottle, and after the two bottles were filled by the addition of water they were allowed to stand another 36 hours. The jelly was almost completely dissolved, so the contents of the two bottles were filtered, first through cheese cloth, then linen, and finally paper. The clear filtrate was divided into several portions which were placed in 2-liter precipitating jars, diluted with an equal volume of water, and the fibrin was precipitated by the cautious addition of 0.3 per ct. acetic acid. At a certain point a flocculent precipitate appeared which quickly settled to the bottom of the jar.

The supernatant liquid was poured off, the precipitate washed with water, dissolved in dilute sodic hydrate (0.05 per ct.), and again precipitated with acetic acid. This process was repeated three times, the final precipitate being washed with alcohol and ether and dried over sulphuric acid in an evacuated desiccator. The fibrin strongly resembled casein in all stages of its preparation except in its extreme sensitiveness to a slight excess of acid or alkali, for, unlike casein, it is readily soluble in weak acetic acid.

The final product was a very fine, light, white powder which gave the following figures upon analysis.

	Per ct.
Moisture	1.84
Ash in dry substance	0.03
Carbon in dry substance	51.82
Hydrogen in dry substance	6.90
Nitrogen in dry substance	17.21
Sulphur in dry substance	0.95
Oxygen in dry substance*	23.12

In order to show that the preparation obtained was not a substance or mixture of substances resulting from the hydrolysis of fibrin by the alkali used to dissolve the fibrin, the following experiment was performed.

Some of the fibrin, prepared as above, was dissolved in 0.2 per ct. sodic hydrate, allowed to stand for 24 hours at 37° C., and the

---

* By difference.

fibrin precipitated by means of acetic acid. The precipitate, after being washed with water, alcohol, and ether, was found to give the same figures upon analysis as the original preparation. The filtrate contained soluble nitrogen which could not be precipitated by acids, showing that the hydrolysis had occurred, but that the products of the hydrolysis did not contaminate the final preparation of fibrin.

THE RELATION OF FIBRIN TO BASES.[2]

Combinations with sodium.—One gram of fibrin was found to require 30.7 c.c. of $\frac{N}{50}$ sodium hydroxide to make it neutral to phenolphthalein, or one gram of fibrin combines with 6.14×10^{-4} gram equivalents of sodium to form a compound neutral to phenolphthalein.

One gram of fibrin dissolved in 50 c.c. of $\frac{N}{50}$ sodium hydroxide requires the addition of 42.4 c.c. of $\frac{N}{50}$ HCl to produce the first sign of a precipitate, or one gram of fibrin combines with 1.52×10^{-4} gram equivalents of sodium to form a compound soluble in water.

One gram of fibrin dissolved in 50 c.c. of $\frac{N}{50}$ sodium hydroxide requires the addition of 50 c.c. of $\frac{N}{50}$ HCl to cause the complete precipitation of the fibrin. This proves that fibrin does not form an insoluble salt containing sodium.

Combinations with calcium.—One gram of fibrin was found to require 30.8 c.c. of $\frac{N}{50}$ calcium hydroxide to make it neutral to phenolphthalein, or 1 gram of fibrin combines with 6.16×10^{-4} gram equivalents of calcium to form a compound neutral to phenolphthalein.

One gram of fibrin dissolved in 100 c.c. of $\frac{N}{50}$ calcium hydroxide requires the addition of 77.5 c.c. of $\frac{N}{50}$ HCl to produce the first sign of a precipitate, or one gram of fibrin combines with 4.50×10^{-4} gram equivalents of calcium to form a compound soluble in water.

One gram of fibrin dissolved in 100 c.c. of $\frac{N}{50}$ calcium hydroxide requires the addition of 85.0 c.c. of $\frac{N}{50}$ HCl to cause a complete precipitation of the fibrin, or one gram of fibrin combines with 3.0×10^{-4} gram equivalents of calcium to form a compound insoluble in water. This precipitate is completely dissolved by 5 per ct. solution of sodium chloride.[3]

RELATION OF FIBRIN TO HYDROCHLORIC ACID.

One gram of fibrin dissolved in 50 c.c. of $\frac{N}{50}$ HCl requires the addition of 42.5 c.c. of $\frac{N}{50}$ sodium hydroxide to produce complete

[2] For the details of the technique involved in these studies consult the work of Van Slyke and Bosworth upon the caseinates (*loc. cit.*).

[3] Casein forms a similar compound with calcium, *Jour. Biol. Chem.* 14:231, 1913, and Tech. Bull. No. 26, of this Station.

precipitation, or one gram of fibrin combines with 1.50×10^{-4} gram equivalents of hydrochloric acid to form a compound insoluble in water.

EFFECT OF CARBON DIOXIDE ON SOLUTIONS OF FIBRINATES.

Fibrin is not precipitated by running a stream of carbon dioxide into solutions of sodium, potassium, or ammonium fibrinates while solutions of calcium or barium fibrinates give precipitates of acid fibrinates. Solutions of calcium fibrinate clot upon being exposed to the air, due to the absorption of carbon dioxide.

Fibrin, unlike casein, does not decompose calcium carbonate when these two substances are triturated together in the presence of water.

MOLECULAR WEIGHT OF FIBRIN.

From the sulphur content we have,

Mol. Wt.$= N [\frac{32.07}{0.95} \times 100] = N 3375.7.$ If $N = 2$, then Mol. Wt.$= 6751.4.$

From the sodium fibrinate containing one equivalent of base we have,

$$\text{Mol. Wt.} = \frac{1}{1.5 \times 10^{-4}} = 6666.6.$$

TECHNICAL BULLETIN No. 42. MAY, 1915.

New York Agricultural Experiment Station.

GENEVA, N. Y.

THE TREE CRICKETS OF NEW YORK: LIFE HISTORY AND BIONOMICS.

BENTLEY B. FULTON.

PUBLISHED BY THE DEPARTMENT OF AGRICULTURE.

THE TREE CRICKETS OF NEW YORK: LIFE HISTORY AND BIONOMICS.

BENTLEY B. FULTON.

SUMMARY

Most of the tree crickets found within the boundaries of New York State are generally distributed over the eastern United States, while some range as far as the Pacific Coast states. They include seven species of the genus *Œcanthus* and one species of the genus *Neoxabea.*

The interest in these insects centers chiefly about their remarkable reproductive structures and instincts and their peculiar oviposition habits. The song of the male, which serves to attract the female, is produced by a minute rasp on the under side of the forewing which is scraped by a structure on the inner edge of the opposite wing. In producing the sound the wings are raised at right angles to the body and vibrated rapidly. When the wings are so raised, there is exposed on the metanotum a glandular hollow, the secretion of which is very attractive to the female. The latter climbs over the abdomen and feeds on the gland. The male takes advantage of the position and inserts the barbed capillary tube of a spermatophore into the genital opening of the female and the sperms pass into the seminal receptacle. The spermatophore is formed in a peculiar pouch at the tip of the abdomen, by the hardening of a viscous liquid about a mass of sperms.

The female prepares for oviposition by chewing a small pit in the bark of the plant. The drilling is accomplished by quick downward thrusts of the ovipositor and a slower twisting motion. After the egg is deposited a quantity of mucilaginous substance is discharged into the hole, and with most species the female plugs up the opening with chewed bark or excrement.

The species of *Œcanthus* can be divided into three natural groups according to their morphology, coloration and habits. The first group includes *niveus, angustipennis,* and *exclamationis.* These are pale whitish crickets, very similar in appearance. All have an intermittent song and deposit their eggs singly in the bark of trees and bushes. The song of *niveus* consists of a series of clear, rhythmically repeated whistles with a pitch about C, two octaves above middle C. The number of notes varies from about sixty to over a hundred and fifty per minute according to the temperature. The songs of *angustipennis* and *exclamationis* are non-rhythmical and have longer notes and rests, generally not more than fifteen per minute.

The second group of species includes *quadripunctatus, nigricornis* and *pini.* The song of these crickets is a continuous, shrill, high-pitched whistle. All three species place their eggs in rows in the pith of small stalks or twigs. *Quadripunctatus* deposits its eggs in loose rows in small or medium sized weeds. The eggs of *nigricornis* are placed in compact rows in large weeds or in the twigs of trees and bushes. *Pini* oviposits only in pine twigs, so far as known.

The third type is represented by a single species, *latipennis.* The song of this cricket is a loud, clear whistle with a musical, ringing quality. The female oviposits in grape vines and weed stalks. A single hole in the outer woody layer of the stem serves for the deposition of from four to twelve eggs, which are placed side by side in the pith in two groups, one above and one below the opening.

Neoxabea bipunctata is closely related to crickets of the genus Œcanthus and was formerly classified with them. Although generally distributed over the eastern half of the United States, it does not extend very far north and is known to occur in this State only in the southeast corner. It has been collected from oak, willow and grape vines, and deposits its eggs singly in the bark or cambium.

INTRODUCTION.

In our studies dealing with the economic aspects of tree crickets, we extended our observations to several species which do not affect cultivated crops, considering especially certain interesting habits which have little or no bearing on the horticultural importance of the insects as a group. These accumulated data have been brought together in this bulletin. The species discussed include only those which inhabit the eastern United States. Several other species of Œcanthus occur in the southwestern states, but living specimens of these could not be obtained and therefore they were not considered in this investigation.

It seemed advisable to include in the present work some material that has already been presented, at least briefly, in Bulletin No. 388, and a few parts, such as technical descriptions, have been copied literally.

All descriptions and measurements are taken from living or recently killed insects, which may differ considerably from dried and shriveled specimens.

GENERAL CHARACTERISTICS.

The term " tree cricket " is commonly applied to any of the tree- or bush-inhabiting insects of the family Gryllidæ. The commonest and best known of these, however, belong to the genus *Œcanthus*, of which there are about twenty-seven known species. According to Kirby[31]* sixteen of these are found on the American continent, while, of the remaining species, one occurs in Europe and the others inhabit various parts of Africa, Asia and the East Indies, with one extending as a rare species into Australia. All members of the sub-family *Œcanthinœ* known to exist in our fauna are included in this genus, except one species which has been placed in a separate genus, *Neoxabea*. These insects are among the most specialized of the Orthoptera and they are characterized by their slender shape, pale color and arboreal habits. From all other members of the Gryllidæ, the *Œcanthinœ* can be readily distinguished by the following characters: The tarsi are three-jointed, and the second segment is small and compressed. All the legs are very slender and in *Œcanthus* are armed with two rows of minute teeth interspersed with a number of delicate spines. The males have very broad and flat tegmina which lie in a horizontal position over the abdomen, while those of the female are narrow and wrapped closely about the body. The two basal segments of the long, filiform antennæ are generally ornamented with peculiar and distinctive black markings which are fairly constant and serve as convenient characters for separating the different species.

The various species of *Œcanthus* in this faunal region seem to fall naturally into three groups as indicated by their morphology, coloration and habits. These groups are typified by the species (1) *niveus*, (2) *nigricornis* and (3) *latipennis*. The first group includes those which live mostly among woody plants, either bushes or trees, and deposit their eggs singly in various places in the bark of the plants. The second group comprises those species which deposit their eggs not in the bark but in the inner pith cavity of small stalks or twigs, making a separate puncture for each egg and grouping them more or less closely in long rows. The third type is represented here by only one species, which places several eggs in the central pith through the same opening in the outer woody layer of the stem, arranging them in two groups, one above and one below the opening. The structural and color characters separating the three groups are indicated in the following key to the species that occur in this region.

---

*Numbers in ( ) refer to Bibliography, pp. 46 and 47.

KEY TO THE SPECIES OF *Œcanthus* FOUND IN EASTERN NORTH AMERICA.

A Basal segment of antennæ with a swelling on the front and inner side. First and second segments each with a single black mark.

B Basal antennal segment with a round black spot. (Fig. 1, a)
niveus De Geer

BB Basal antennal segment with a J-shaped black mark. (Fig. 1, b) *angustipennis* Fitch

BBB Basal antennal segment with a straight club-shaped black mark. (Fig. 1, c)
exclamationis Davis

AA Basal antennal segment without a swelling on the front and inner side. First and second antennal segments each with two black marks or entirely black. Tegmina of male 5 mm. or less in width.

B Head and thorax pale yellowish-green or black or marked with both colors.

C First antennal segment with a narrow black line along inner edge and a black spot near the distal end. Body entirely pale yellowish-green. (Fig. 1, d)
quadripunctatus Beut.

CC First antennal segment with black markings similar to above, but broader and usually confluent, sometimes covering the whole segment. Head and thorax often with three longitudinal black stripes: ventral side of abdomen always solid black in life. (Fig 1, f, g)
nigricornis Walker

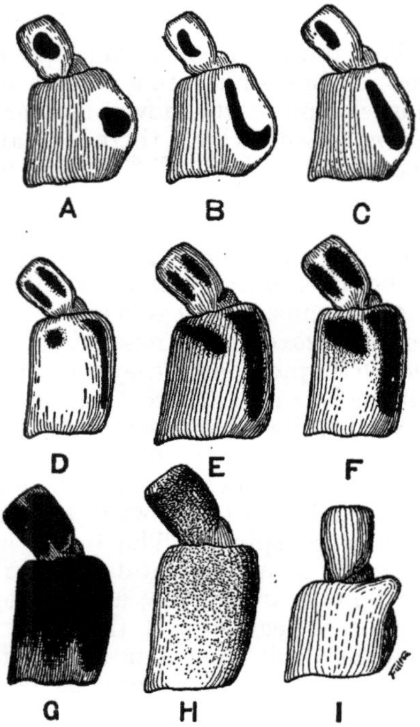

FIG. 1.— BASAL ANTENNAL SEGMENTS OF TREE CRICKETS.

a, *Œcanthus niveus*; b, *Œ. angustipennis*; c, *Œ. exclamationis* ; d, *Œ. quadripunctatus*; e, *Œ. pini*; f, *Œ. nigricornis*; light form; g, *Œ. nigricornis*, dark form; h, *Œ. latipennis*; i, *Neoxabea bibunctata*.

BB Head, thorax and antennæ reddish brown. Wings in life with conspicuous green veins. Marks on basal antennal segment broad but seldom confluent. (Fig. 1, e) *pini* Beut.

AAA Basal antennal segment without a swelling on the front and inner side. Basal portion of antenna red, unmarked with black. Tegmina of male about 8 mm. wide. (Fig. 1, h) *latipennis* Riley

GENERAL DESCRIPTIONS OF LIFE STAGES OF ŒCANTHUS.

Egg.— The eggs of the species belonging to the genus *Œcanthus* are elongate, cylindrical and slightly curved. At the time of deposition they are semi-transparent, but later they grow more opaque and become slightly swollen. The cephalic end possesses a whitish opaque cap, which is covered with minute projections, arranged in spiral rows after the fashion of the scales of a pine cone. At the extreme base of the cap, only shallow rhombic depressions occur, but above the first few rows short projections appear between the indentations and gradually increase in size in each successive row. The entire surface of the egg, exclusive of the cap, is etched with what appear to be very minute cross-hatched scratches. The number and size of the spicules vary considerably with different species and are useful characters for distinguishing the eggs of certain of the crickets.

Nymph.— The newly hatched nymphs are very pale and delicate. After feeding, the size of the abdomen is increased and the contents of the alimentary canal show through the body wall and give the insect a darker color and more robust appearance. The appearance of the nymphs and the characters distinguishing the different instars are shown in Plate I and Fig. 2. The first instar is characterized by having only thirty-four segments in the antennæ while the sides of the meso- and metanotum do not project, and they extend downward scarcely as far as do the sides of the pronotum. The second instar differs from the first by a slight downward extension of the sides of the meso- and metanotum. The edges are free and lap over the pleura. The antennæ have about double the number of segments of the first stage. The third instar is characterized by a still further prolongation of the sides of the meso- and metanotum so that they appear as distinct and rudimentary wing pads. The antennal segments are increased to nearly one hundred. On the fourth instar the wing pads are folded back over the thorax and reach to the second abdominal segment. In the fifth instar they reach to the third or fourth segment.

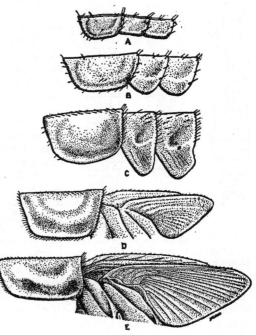

Fig. 2.—Thoracic Segments of the Five Nymphal Instars.

The differences between the three groups of species are plainly evident in the first instar (Fig. 3). In the *niveus* type the nymphs are white with a few fine black marks on the head and thorax; in the *nigricornis* type they are pale greenish yellow with a slight dorsal infuscation and a pale median line; and the nymphs of *latipennis* are white with a purplish red dorsal area and a pale median line.

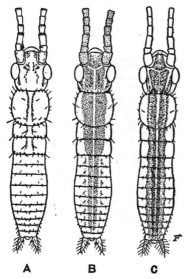

Adult (Plate I, fig. 6).— Body slender. Long axis of head in life nearly horizontal. Pronotum slightly elongated; disk narrowed in front; sides nearly vertical at hind edge, flaring outward in front part. Abdomen composed of ten segments, including the last which bears the cerci and anus. First segment much reduced ventrally. Antennæ filiform; over twice the length of body. Legs slender; hind femora a little thickened. Hind tibiæ with a double row of teeth on posterior side, intermixed on distal half with longer spines; with three pairs of spurs at the tip. Anterior tibiæ with a tympanum near the base.

FIG. 3.—NYMPHS OF THE FIRST INSTAR, ILLUSTRATING THE THREE TYPES OF *Œcanthus*.

a, *niveus*; b, *nigricornis*; c, *latipennis*.

The hind wings of both sexes are folded and generally extend beyond the tip of the fore wings.

Males have the fore wings much broader than body. A longitudinal fold occurs about one-third way from the costal margin; the inner two-thirds lies horizontally over the back and the remainder is deflexed toward the sides.

The fore wings of the female are regularly reticulated, much narrower than those of the male, and are wrapped closely around the body. A longitudinal bend of about ninety degrees is located about half way between the two margins.

HATCHING.

The time of hatching seems to be fairly uniform for the four species of which data on this process were obtained, namely *Œ. niveus, angustipennis, nigricornis* and *quadripunctatus*. For average years in western New York it begins from the tenth to the middle of June and the insects may continue to appear until about the twentieth of that month. Farther south the time of hatching is correspondingly earlier.

The process of hatching is best explained by the accompanying figure (Fig. 4). The egg becomes slightly swollen, preceding the period of hatching, apparently due to an increase in internal pressure. The end of the cap breaks off, and the embryo slips out (Fig. 4, b). It retains its embryonic form until about half way out, then the thorax becomes strongly arched upward (Fig. 4, c, d). The nymph continues to work itself out by muscular contractions of the abdomen and by moving the body up and down and from side to side. A delicate membrane projects from the egg puncture and clings to the body, making it difficult for the nymph to free the hind legs and antennæ. The latter are finally grasped by the mouth parts and worked out

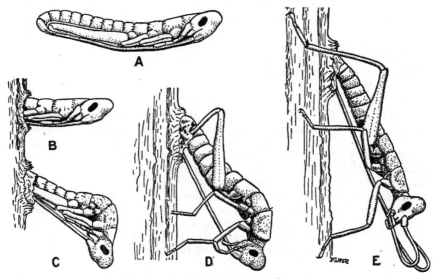

FIG. 4.— HATCHING OF TREE CRICKETS.

a, Position of embryo in egg; b, c, d, e, Successive stages in emergence of nymph. Drawings made from Œ. quadripunctatus.

by movements of the head (Fig. 4, e). Within about ten or twelve minutes the nymph is free and crawls upward on the plant. A prominent watery bump on the head, which formerly filled out the end of the egg, gradually diminishes in size and within half an hour is not apparent. Some nymphs never succeed in freeing themselves from the egg and these are often devoured by their more fortunate companions.

In all cases observed the ventral side of the embryo lies next to the outer curve of the egg, thus if the egg curves downward in the plant, the embryo will emerge ventral side up. The eggs of Œ. quadripunctatus generally turn upward in the stalk, while those of

Œ. nigricornis turn downward in nearly all cases. The eggs of the other species turn downward much more often than upward.

MOLTING.

When a nymph prepares to molt, it first fastens its claws firmly in the bark or in the tissues of a leaf, extends the antennæ backward and arches up the back. The skin splits along the dorsal median line of the head and thorax. The head is bent down and the thorax works out through the split. The fore and middle legs are pulled out and exercised, while the palpi and antennæ are still held in the skin. The hind legs are pulled upward and forward. The antennæ are partly pulled out by straightening the body, and then they are grasped by the mouth and worked out in the same manner as noted in the process of hatching. When the hind legs are free the nymph grasps the support and pulls out the hind part of the abdomen. Later the skin is eaten by the insect if in the meantime the discarded remnant has not been consumed by some other cricket.

FEEDING HABITS.

The food of the crickets consists of a great variety of materials of both plant and animal origin, which are soft enough to be masticated by their comparatively weak mandibles. In the breeding cages the crickets were fed almost entirely on plant lice, which were readily available and on which the crickets seemed to thrive well. Any cricket which was injured or in any way unable to defend itself was generally devoured by its companions. They also ate numerous small holes in raspberry leaves, and chewed at the cambium layer of apple branches in places where it was exposed by a cut. They ate the anthers and other floral parts of the wild carrot (*Daucus carota*) and also chewed out patches of the outer tissue of the stalks. Tree crickets confined on branches with ripe peaches and plums ate round holes in the sides large enough to insert their heads in order that they might feed on the inner pulp. They would also subsist on slices of ripe apples placed in the breeding cages, but at no time did they touch the whole fruit that was sound, apparently being unable to penetrate the outer skin. Tree crickets in the breeding cages readily fed on San José scale on badly infested apple branches. They devoured both the protective covering or scale and the insect itself. One nymph of *niveus* in the fourth stage, which was confined to this diet, ate on an average about four hundred scales during a night, counting both mature and immature insects, and on one night devoured over nine hundred individuals of this coccid.

For further information on the natural feeding habits of the tree crickets, the crops of a number of specimens of *niveus* and *angustipennis*, from apple and *nigricornis* from raspberry, were dissected out and the contents examined under a microscope. In most specimens a large part of the diet consisted of plant tissue in which could be recognized chlorophyll-bearing cells, leaf hairs, vascular tissue, mycelia and spores of various fungi. Crops from *niveus* and *angustipennis* usually contained a small amount of insect material, and in some specimens this was the main constituent. In many cases, it was possible to detect parts of aphids, scale insects and what appeared to be the cast skins of the crickets although these materials were usually much broken up and barely recognizable. Two crops contained a few lepidopterous wing scales and in another the leg and wing of a small hymenopterous insect were detected. Specimens of *nigricornis* always appeared to feed more extensively on plant tissue, although in some specimens a few insect remains could be distinguished.

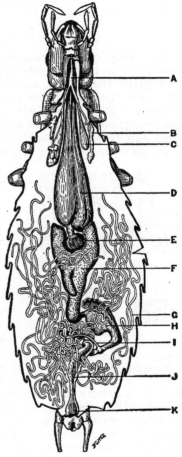

DIGESTIVE ORGANS.

The digestive organs of tree crickets have some peculiarities worthy of notice. The crop (Fig. 5, d) is large and thin walled and reaches to about the middle of the body; the proventriculus (Fig. 5, e) is spherical and lies between the two anterior lobes of the stomach. The ventriculus or stomach is divided by a constriction into two parts; the anterior, thick walled portion (Fig. 5, f) bears two cœca at the anterior end and one on the right side about the middle; the posterior, thin walled portion (Fig. 5, h) bears a cœcum at the anterior end; the posterior end suddenly narrows down before passing into the hind intestine. At this point a small duct enters (Fig. 5, i) which arises at the junction of the numerous, long

FIG. 5.— DIGESTIVE SYSTEM OF TREE CRICKETS.

a, Common salivary duct; b, salivary gland; c, salivary reservoir; d, crop; e, proventriculus; f, fore part of ventriculus; g, malpighian tube; h, hind part of ventriculus; i, common duct from malpighian tubes; j, hind intestine; k, rectum.

malpighian tubes (Fig. 5, g). The proctodæum or hind intestine is short and has only one bend in it (Fig. 5, j).

The pair of salivary glands lie in the thorax (Fig. 5, b). The ducts from each receive a branch from the associated thin walled reservoir (Fig. 5, c) and unite under the œsophagus into a common salivary duct (Fig. 5, a) which opens under the hypopharynx.

MUSICAL ORGANS AND SONG PRODUCTION.

The songs of tree crickets form a considerable part of the insect sounds to be heard in late summer and autumn. The males generally place themselves in some hidden retreat among the leaves, with only their long antennæ projecting to warn them of approaching dangers. They stop singing at the slightest jar of the ground or movement of the plants in which they are located, but at night a strong light can be thrown on them without appreciably disturbing them.

In preparing to sing the male raises the fore wings or tegmina perpendicular to the body. This movement automatically unfolds them so that the inner portion, which normally lies horizontally over the back, and the inflexed outer portion, come to lie in the same plane when the wings are raised. The sound is produced by the fore wings vibrating rapidly in a transverse direction, so that the overlapping inner portions rub against each other.

FIG. 6.— LEFT WING OF *Œ. latipennis,* SHOWING RASP (A) AND SCRAPER (B).

The mechanism which produces the sound is found near the base of the wing, the broad expanded distal part serving as a resonator to increase the volume of sound. A short but prominent transverse vein, about one-fourth way from the base, is modified beneath to form a minute filiform rasp (Fig. 6, a). It is from one to one and a half millimeters long, according to species, and bears from twenty to fifty short teeth which are inclined slightly toward the opposite wing. In all of many specimens examined the right wing laps over the left. The latter has a fine thickened ridge along the inner edge just opposite the file (Fig. 6, b). This scrapes against the teeth of the file on the right wing and thus produces the sound vibrations. The underside of the left wing has a file practically identical to the other but this is apparently seldom, if ever, used. In fact it is doubtful if it ever could be used, for the scraper on the right wing is imperfectly formed and in most cases at least lacks the chitinous ridge along the edge.

MATING HABITS.

The peculiar mating habits of the Gryllidæ have aroused the interest of a number of entomologists both in this country and in

Europe. The first careful and detailed study of the subject was made by Lespès[2, 3] in 1855. In recent years their activities in this respect have been carefully considered by Hancock,[29, 42] Houghton,[34, 35] Baumgartner[40] and Jensen[36, 43] in this country and by Cholodkovsky,[41] Boldyrev[46] and Engelhardt[48] in Russia. My own observations show that the mating habits of *Œcanthus* are similar in many respects to those of *Gryllus* and *Nemobius* as studied by Baumgartner.

The male calls vigorously until a female comes near him, in which case he sidles toward her, without any cessation in his singing and keeping his head in an opposed direction. If he succeeds in attracting her attention she climbs over his body and begins to feed on the secretion of a glandular cavity on the metanotum, which is described by Hancock as an "alluring gland." The male stands with legs spread wide apart and the tegmina are held at an angle of about 45 degrees above the abdomen (Plate II, Fig. 1). His body sways and twitches considerably and the hind wings, which lie folded along the back have a peculiar jerky movement whenever the female bites at the gland. The antennæ are waved about wildly and are often thrown back so as to cross and rub against those of the female. After about half an hour, the male reaches back with his abdomen and the female bends her abdomen downward. This

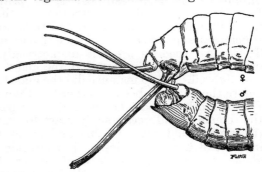

FIG. 7.— POSITION OF BODIES OF MALE AND FEMALE WHILE TRANSFERRING SPERMATOPHORE.

enables him to slip a pair of small, laterally compressed blades into the notch at the tip of the female's subgenital plate. (Fig. 7.) At this time the barbed capillary tube of a spermatophore is pushed into the vagina and when the tip of the abdomen is withdrawn the bulb of the spermatophore is drawn out of its pocket in the male and remains fastened to the female. The latter does not leave immediately but continues to partake of the secretion of the gland for a half hour or more. She finally crawls away to some secluded spot and arches up her back, bringing the tip of the abdomen forward beneath and pulls off the spermatophore with her mouth. She straightens out again and proceeds to eat the capsule in a leisurely way, after which she doubles up again and works at the ovipositor with her mouth, as if endeavoring to clean it.

In fastening the spermatophore to the female, the male places his cerci on opposite sides of the ovipositor, and they appear to guide him in striking the proper opening. A male from which both cerci had been removed at the base, was observed trying to copulate. When he succeeded in striking the base of the ovipositor, the female turned the tip of the abdomen down as usual. The male then passed the pair of chitinous blades up and down the ovipositor but was unable to strike the opening at the base and after several attempts the pair became separated.

METANOTAL GLAND.

The exact function of the gland on the metanotum of the male has been a matter of doubt. Hancock[29] first described it as an "alluring gland," claiming that it served to attract and hold the attention of the female until copulation could take place. Boldyrev[46] and Engelhardt[48] of Moscow, Russia, have concluded from observations on the European species, Œ. pellucens Scop., that its chief function is to hold the attention of the female after the spermatophore has been attached, and thus prevent her from devouring it before the sperms have had time to pass into the seminal receptacle. These two theories are not diametrically opposed, and from my own observations it seems that the gland may serve both purposes equally well. In no instance did the male transfer the spermatophore before the pair had been together less than a quarter or half an hour. During both periods, before and after passing the spermatophore, the female would occasionally start to crawl away. At such times the male would begin to sing and follow after her, placing himself in front of her and taking such a position as to expose the alluring gland. Usually the female would return, but sometimes she would leave to stay, even before the spermatophore had been fastened to her. Considering the length of time for the first part of the mating process, and the willingness with which the female departs from the male, it seems doubtful if the male, without the help of the gland, could hold the attention of the female long enough to place the spermatophore in the genital opening. On the other hand, the female devours the spermatophore very soon after leaving the male, and if the insects separated immediately after the spermatophore was given over, only a small part of the sperms would have time to flow into the seminal receptacle.

In this connection it is of interest to note that a female nymph of the fifth instar was at one time observed eagerly feeding on the gland of a mature male. This behavior of the nymph would indicate that the mature females probably have a taste for the secretion of the metanotal gland independent of the act of copulation with the male.

Externally the metanotal gland appears as a triangular depression situated in the center of a broad rounded disk, between the bases of the hind pair of wings (Fig. 8 and Fig. 10, a). Most of the surface of the disk is thinly clothed with long delicate hairs. The triangular hollow, the apex of which is directed anteriorly, is densely clothed with shorter hairs. On the sides are two paired rows of

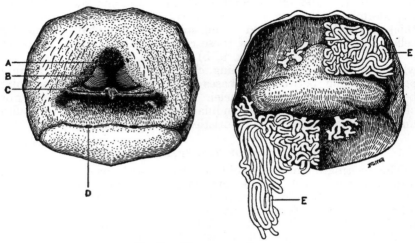

FIG. 8.— METANOTAL GLAND.

Left, dorsal view; right, ventral view of internal structures; a, Opening of anterior tubular gland; b, anterior tuft of hairs; c, posterior tuft; d, opening of posterior tubular gland; e, tubular glands.

larger curved bristles. The hairs of the anterior row are distinctly separated and do not reach to the median line (Fig. 8, b). The pos-

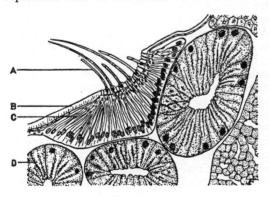

FIG. 9.— METANOTAL GLAND.

Section through anterior tuft of hairs; a, hair; b. cuticle; c, hypodermis cells; d, tubular gland.

terior row forms a dense brush which reaches to the median line, and at the tip the bristles curve sharply forward and then point backward. (Fig. 8, c.) Two pairs of openings can be discerned with difficulty; the anterior pair on the sides of the depression just above the anterior ends of the first row of bristles (Fig. 8, a) and the second pair in the transverse suture at the extreme posterior edge of the depression (Fig. 8, d).

By dissection it can be shown that these paired openings lead from much branched internal glands (Fig. 8, e, and Fig. 9, d).

These glands occupy most of the space immediately beneath the metanotum and many long tubes from the posterior pair extend back into the abdominal cavity. Stained sections show that the hypodermis underlying the hairy parts of the gland is unusually thick and that it is composed of long spindle- or club-shaped glandular cells (Fig. 9, c). Certain of these cells pass through minute pores in the chitinous cuticula (Fig. 9, b) and into the hairs (Fig. 9, a), which contain a cavity running clear to the tip. Other minute pores open out between the bases of the hairs. The larger hairs bear several spindle-shaped nuclei. All of the hairs on the interior of the cavity appear to be glandular but the largest cells are found under the two paired tufts of bristles as described above.

DESCRIPTION OF THE SPERMATOPHORE.

The spermatophore is ovoid in shape, with one end tapering into a conical neck which terminates in a long capillary tube (Fig. 11, v). The main body varies somewhat in size with the different species, from about .6 mm. long and .5 mm. wide in *Œ. angustipennis* to about 1.1 mm. long and .9 mm. wide in *Œ. latipennis*. The wall consists of two translucent coats. The outer (Fig. 11, m) is white and rather easily crushed and forms only a thin film at the tube end, but gradually becomes very thick at the opposite end; and it is this that gives the spermatophore its oval shape. The inner coat (Fig. 11, k) which forms a nearly spherical body, is very hard and firm, of a pale brown color, darker near the inner cavity and has a uniform thickness of about one-sixth the entire diameter of the spermatophore. The inner cavity (Fig. 11, j) is filled with sperms at first but if the spermatophore is removed from the female after mating, it is nearly or entirely empty. The tube is nearly one and one-half millimeters long and is bent in the form of a hook, with the distal half straight and pointing back over the main body (Fig. 11, e). It has an outside diameter of about .03 mm. near the base and gradually tapers to a very fine point. Just distal to the bend of the tube is a thin blade-like barb. The cavity of the body of the spermatophore is continuous with that of the tube.

When the spermatophore is fastened to the female the sperms start to pass out rapidly through the tube. I have watched this process in the spermatophore of *Gryllus pennsylvanicus* but not in the spermatophore of *Œcanthus* species. The spermatophore was removed from the male and placed on a slide with a little water around the tip of the tube but not around the bulb. A granular looking substance came out slowly at first. Then the spermatozoa began suddenly to shoot out rapidly through the tube, which was so small as to allow only a few to pass through side by side After about fifteen minutes the sperms were all out and these were followed by a brownish substance which was discharged for a short time. The bulb of the spermatophore did not collapse and seemed to retain its normal size after the loss of its contents.

MALE REPRODUCTIVE ORGANS.

The testes are irregularly oval, pale yellow bodies located in the anterior part of the abdomen (Fig. 10, b.). The seminiferous tubules, of which each one is made up, converge into a sinus near the anterior end. The vas deferens is a minute tube which runs from the sinus back between the tubules and emerges near the posterior end of the organ (Fig. 10, c, and Fig. 11, d). It continues along the side of the body to the posterior extremity where it turns forward and becomes much enlarged and coiled, forming an oval body surrounded by yellow fatty tissue (Fig. 10, g, and Fig. 11, c, g). This enlarged portion is filled with sperms and no doubt serves the function of a seminal vesicle. The short remaining portion of the vas deferens is straight and of larger diameter, and continues anteriorly to join the large sinus into which hundreds of glandular tubules empty (Fig. 10, d, and Fig. 11, b). These tubules are filled with a homogeneous substance which probably forms the wall of the spermatophore. From the sinus a large common duct (Fig. 10, e) leads to the spermatophore mold at the end of the abdomen (Fig. 10, h). That this duct is not a true ejaculatory duct,

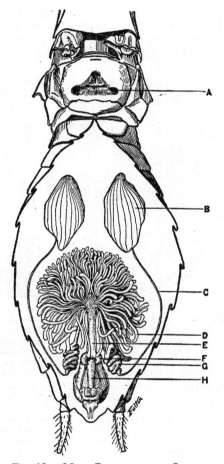

FIG. 10.— MALE REPRODUCTIVE ORGANS. a, Metanotal gland; b, testis; c, vas deferens; d, glandular tubule; e, common duct; f, accessory pouch; g, enlarged portion of vas deferens; h, spermatophore mold.

either in origin or function, has been pointed out by Baumgartner in his work on *Gryllus*. It arises as an invagination from the outside and is not used as a sperm carrier during copulation.

The spermatophore mold lies in the ninth abdominal segment, the sternum of which forms the large subgenital plate. The part of the wall of the mold which contains the body of the spermatophore consists of two pliable flaps which meet on the median line dorsally and ventrally. These are collapsed when empty but conform to the bulb of the spermatophore when one is present (Fig. 11, l). At the anterior end a groove or out-pocket of the mold starts in and passes first dorsally and then posteriorly (Fig. 11, f). This groove is

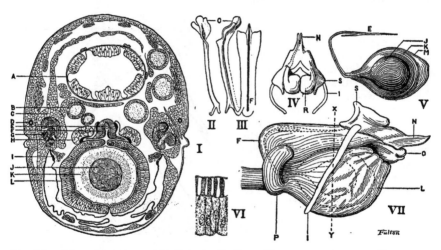

Fig. 11.— Male Reproductive Organs: I, Cross Section of Abdomen Through Spermatophore Mold. II, Lateral View of Tri-lobed, Paddle-shaped Structure. III, Ventral View of Same and Groove of Spermatophore Mold. IV, Ventral View of Copulatory Blades and Basal Structure. V, Spermatophore. VI, Portion of Epithelium Surrounding Spermatophore. VII, Lateral View of Spermatophore Mold.

a, Rectum; b, glandular tubule; c, fat-tissue; d, vas deferens; e, spermatophore tube; f, groove of spermatophore mold; g, enlarged portion of vas deferens; h, muscle; i, chitinous supporting rod; j, sperm cavity; k, inner layer of spermatophore wall; l, wall of spermatophore mold; m, outer layer of spermatophore wall; n, copulatory blades; o, paired paddle-shaped structure; p, accessory pouch; r, paired cavity and tube running into copulatory blade; s, basal structure and point of attachment; x-y, location of accompanying cross section.

supported by chitinous walls and terminates in a sharp point, and just before the tip it bears a pair of horizontal wings which together resemble a fish-tail in shape. The groove holds the tube of the spermatophore (Fig. 11, e). On each side at the level of the groove is a movable, spoon-shaped piece, which is three-lobed at the tip and connects with the groove support at the base (Fig. 11, o). Dorsal and posterior to the end of the groove is a pair of laterally compressed, blade-like appendages which lie close together and terminate in a slightly hooked point (Fig. 11, n). These blades are inserted into the genital opening of the female when the spermatophore is transferred. At the base they widen out and are united on the median l'ne. Still farther anteriorly the heavily chitinized part divides into two large basal structures, which are connected with the posterior edge of the lateral walls of the ninth segment and constitute the main supporting structure of the spermatophore mold (Fig. 11, s). Between these structures near the median line is a pair of closed cavities, each with a tube extending posteriorly into the corresponding blade-like appendage to near its tip (Fig. 11, r). If these tubes have an opening in the blade it is very minute and could not be detected. A pair of curved chitinous rods are attached to the basal structure and extend obliquely around the mold to nearly the ventral median line (Fig. 11, i) and help support the soft flaps covering the body of the spermatophore. A pair of elongated pouches arise at the junction of the spermatophore mold and the common duct, and extend dorsally along the side of the groove (Fig. 11, p). Their function is not known.

Stained sections show that the lining walls of the mold including the groove are formed of a rather thick layer of columnar epithelium (Fig. 11, VI). This is of ectodermic origin and the side next to the inner cavity is bordered by a thin layer of cuticle which becomes, thicker in parts of the groove. The cells show a peculiar, sharply demarked differentiation of the ends toward the inner cavity, where about one-third of the length is much narrower and rod-shaped, and the intervening spaces are wide. This narrow part is darkly and uniformly stained while the wide nucleated portion has a granular appearance.

FORMATION OF THE SPERMATOPHORE.

The exact process of spermatophore formation is as yet only a matter of conjecture. It seems very probable, however, that a mass of sperms which have collected in the enlarged portion of the vas deferens move out into the mold and are later surrounded by a quantity of secretion from the glandular tubules which hardens around them and in some way forms the spermatophore. The tube is probably formed in the groove, but why it develops as a tube and not as a solid rod is a mystery.

A new spermatophore is usually formed soon after the last one is removed. In mating it was observed that in about ten minutes after the spermatophore is passed to the female and while the latter is still feeding at the metanotal gland, the collapsed mold begins to swell out and in another minute or so is fully distended and remains so. In one case a male was examined about half an hour after the mold became distended. The mold contained a white, globular mass of sperms with a long attenuated thread at one end. It appeared to be enclosed in a delicate membrane although this could not be distinguished. It was surrounded by a clear viscid liquid of about the consistency of egg albumen. I removed the sperm mass with forceps and the liquid began to congeal. A small drop on the point of a needle could be pulled out into a very fine thread which would harden very quickly. The hardening continued even after the mass was placed under water and resulted in a hard substance like the wall of the spermatophore.

Another male which had disposed of a spermatophore was examined in an hour and three-quarters after the mold had become distended and it was found to contain a fully formed and hardened spermatophore.

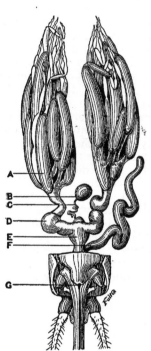

FIG. 12.— FEMALE REPRODUC-
TIVE ORGANS.

a, Ovary; b, seminal receptacle; c, duct leading to seminal receptacle; d, oviduct; e, vagina; f, accessory gland; g, copulatory opening.

FEMALE REPRODUCTIVE ORGANS.

In order to follow the course of the sperms after the deposition of the spermatophore it is necessary to understand a little of the anatomy of the female reproductive system. The ovaries occupy a large portion of the central part of the abdominal cavity. They are made up of a large number of separate egg tubes, each of which contains several ovarian follicles in various stages of development (Fig. 12, a). The tubes all converge into the broad, thin-walled oviduct (Fig. 12, d) and these unite to form the vagina (Fig. 12, e) which continues as a straight tube to the ovipositor. A single tubular accessory gland (Fig. 12, f) opens above the vagina at the base of two short rods which are between and hidden by the four large rods of the ovipositor. This gland probably furnishes the mucilaginous substance exuded at the time of oviposition. The notch at the end of the subgenital plate opens into the ventral side of the vagina (Fig. 12, g). Just anterior to this on the dorsal wall of the vagina is a hard plate, which is slightly hollow on the inner side and bears a small hole in the center. This hole passes through and opens on the dorsal side into a small convoluted duct which leads to an ovoid seminal receptacle lying just anterior to the junction of the two oviducts (Fig. 12, b, c).

A female bearing a spermatophore was killed and dissected. The tube was inserted in the notch at the end of the subgenital p'ate, and the point was directed into the hole in the plate on the dorsal wall of the vagina. It would appear that the sperms discharging through the tube would thus be directed into the duct of the seminal receptacle.

OVIPOSITION.

For this operation the female selects a suitable spot on the plant and prepares to oviposit by first chewing a small hole in the bark. Upon the completion of the cavity she then walks forward a little, arches her back so as to bring the ovipositor about perpendicular to the branch and begins moving it up and down until she strikes the hole. She then starts to dril! by giving the ovipositor quick thrusts and at the same time slowly turning it around by twisting the abdomen thirty or forty degrees to each side (Plate II, Fig. 2). It takes from five to twenty-five minutes according to the resistance of the wood, to force the ovipositor to its base the first time. After this it is pulled nearly out and drilled in again several times, each operation requiring from one to five minutes or more. When the hole is sufficiently reamed out and the ovipositor drilled in for the last time, the female rests a moment and then arches up the middle of the abdomen and forces an egg down into the ovipositor. The body quivers and shakes a little during the operation, which lasts about twenty seconds, and then the ovipositor is slowly withdrawn, twisting from side to side. When only the tip remains in the hole the female pauses and discharges a small quantity of mucilaginous substance which fills up the opening. She then removes the ovipositor and plugs up the mouth of the hole, using such material as the instincts of the various species determine. She spends several minutes packing it in and smoothing it out so that the wound is neatly capped. The whole process of oviposition, from chewing the bark to sealing the wound may require from a quarter to nearly a whole hour.

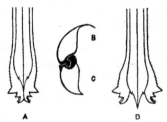

FIG. 13.— OVIPOSITOR.

a, Tip with rods in normal position; b, c, cross section of upper and lower rods on one side; d, position of rods when ventral pair are thrust downward.

The ovipositor is made up of two pairs of long chitinous rods which fit together in such a way as to make a compact organ nearly circular in cross section. It was observed that, in drilling, the upper pair of rods remain stationary except for the twisting movement while the lower pair slide up and down with each thrust. The upper and lower rod on each side can not be separated readily and in cross section it can be seen that they are firmly united by a tongue of the upper fitting into a groove in the lower (Fig. 13, b, c). At the tip the lower pair are wedge shaped and fit in between the ends of the upper pair, which have four stout spurs on each side. When the lower rods are thrust downward their chisel-shaped points push into the bottom of the hole and at the same time they wedge apart the ends of the upper rods causing them to widen the hole (Fig. 13, d).

The ovipositor first makes its appearance in an incipient stage in the second instar in the form of two small rounded projections on the ventral side of both the eighth and ninth abdominal segments. In the third instar it consists of four backward pointing papillæ — the anterior pair being smaller and lying in contact with the lower surface of the hind pair. In the fourth instar it forms a compact body extending a trifle beyond the tip of the abdomen. The sterna of the eighth and ninth segments are reduced and almost completely taken up by their appendages. In the fifth instar the ovipositor is much longer and bears much resemblance to the adult organ. The sternum of the seventh segment is enlarged and forms the subgenital plate.

KEY TO EGGS AND OVIPOSITION HABITS OF THE TREE CRICKETS.

A Eggs white or only slightly yellowish; deposited singly in the bark or cambium of trees and bushes.
    B Egg cap with long finger-shaped projections. *niveus*
    BB Egg cap with knob-like projections, not much longer than thick.
        angustipennis
    BBB Egg cap with low rounded elevations. *exclamationis*
AA Eggs yellow; deposited in rows, and lie in the pith of stems of herbaceous plants and twigs or small branches of woody plants.
    B Eggs in loose rows, in small herbaceous stems, 5 mm. or less in diameter. Projections of cap, finger- or club-shaped. *quadripunctatus*
    BB Eggs in compact rows, in stems 5 mm. or over in diameter. Projections of cap with length not more than one and a half times the width.
        nigricornis
    BBB Eggs in pine twigs. Projections of cap transversely elliptical in cross section. *pini*
AAA Eggs pale yellow; deposited in two groups in the pith, one cluster on each side of the hole made in the outer woody layer of the stem. The holes may be in rows but in such case are a centimeter or more apart. *latipennis*

ŒCANTHUS NIVEUS De Geer

HISTORICAL NOTES.

This insect was described from a Pennsylvania specimen as early as 1773 by De Geer,[1] who called it *Gryllus niveus*. Its status has been clearly established in systematic literature but in economic writings it has long been confused with *nigricornis* and regarded as the cause of an oviposition injury to raspberries performed by the latter species. Walsh[6] and Riley[7] evidently did not distinguish between the two species for both describe partly black varieties which never occur in *niveus*. Later Riley[10] states that he considers both *nigricornis* and *fasciatus*, which are synonymous, as dark varieties of *niveus*. The uncertainty as to the identity of these species by economic writers has probably led to the subsequent confusion concerning their oviposition habits, which still persists to a less degree.

DISTRIBUTION.

From available records *niveus* is the most generally distributed of the species occurring in the United States, ranging from Massachusetts to the Pacific Coast and from the Province of Ontario on the north to as far as Mexico on the south It is found all over the State of New York with the exception of forested regions in the northeastern part. It has been recorded in literature from the following states: Massachusetts (Faxon), Connecticut (Walden), New Jersey (Davis), Ontario (Walker), Georgia (Allard), Indiana (Blatchley), Illinois (Forbes), Kentucky (Garman), Minnesota (Lugger), Kansas (Tucker), Nebraska (Bruner), Michigan (Allis), Cuernavaca, Morelos, Mexico (Rehn), Cuba (Kirby). From specimens examined we can record its distribution in the following states: Colorado and Utah (Titus), Ohio (Kostir), New Jersey, North Carolina, Connecticut (Amer. Mus.), California (Doane); Maine, one specimen (Patch), Cuba (Cardin). From correspondence we have obtained other records as follows: Texas (Newell), North Carolina (Beutenmüller), California and Washington (Melander).

HABIT.

Œcanthus niveus is a tree- and bush-inhabiting form. It is found most abundantly in apple orchards and is more or less common in plantings of other fruit trees and in raspberry plantations, shrubberies, vines and bushy fence rows. Among forest trees it is less common, although a few can often be heard singing in such places. especially along the edge of a wood. In general this species prefers a cultivated region to a wilderness. However, in orchards that are regularly sprayed with arsenicals the crickets do not become very abundant.

DESCRIPTION OF LIFE STAGES.

Egg (Fig. 14, g).— The egg is about one-ninth of an inch long and from one-sixth to one-fourth as wide. The color is dull white, often with a slight yellowish tinge. The cap (Fig. 14, f) is a little narrower than the main body; its sides are parallel and the end is broadly rounded. In color it is opaque white, but is often stained a reddish color by the bark. The projections on the cap (Fig. 14, e) are long and finger-shaped, having a uniform thickness of about .009 mm. from base to tip, and a length of .020 to .025 mm.

The average measurements of forty specimens of eggs are as follows: length 2.83 mm.; greatest width, .62 mm.; length of cap .51 mm.; width of cap .51 mm.

Nymph.— First instar (Plate I, Fig. 1): Color white. Top of head with two rows of ten to fifteen small bristles, directed anteriorly and each with a small black spot at the base. There is a short black line extending backward from the upper edge of each eye and one or two pairs of brownish transverse spots between the eyes. The pronotum, and sometimes the meso- and metanotum, have a pair of longitudinal brownish stripes situated close to the median line. Basal segment of antenna with a small black spot on the inner side and a brownish spot on the posterior side; second segment with a black transverse line on the inner side; third, fourth, sixth and ninth segments with a narrow black ring at apex; each succeeding segment with faint gray annulation at tip. Hind tibiæ with black spots at the base of the small bristles, especially prominent on the outer and upper sides. Length about 3 mm. Antennæ 6.3 to 7.5 mm.

Second instar (Plate I, Fig. 2): Ground color of abdomen transparent greenish white with two rows of pure white blotches on each side of median line. Basal segment of antenna with a round black spot on the front and inner side, and the second segment with a similar spot on the front side and a transverse dash on inner side. Distal part of antennæ with gray annulations on alternate segments. Length 4.5 to 5 mm. Antennæ 10.7 mm.

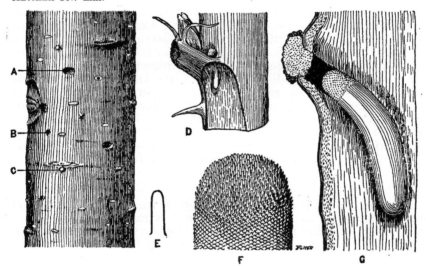

FIG. 14.— *Œcanthus niveus* DE GEER.

a, Egg puncture of previous year healed over, in apple bark; b, recent egg puncture without plug; c, egg puncture with plug (all x 1½); d, egg in raspberry (x 2½); e, projection of egg cap (x 500); f, egg cap (x 50); g, egg in apple bark (x 15).

Third instar (Plate I, Fig. 3): General color greenish white. Abdomen with several rows of irregular opaque white blotches on each side of median line. The brownish markings on the head and thorax are very faint. Black spot on first segment of antenna on a white prominence. Length 6 to 7 mm. Antennæ 13 mm.

Fourth instar (Plate I, Fig. 4): Coloration practically the same as in the preceding stage. Length 8.5 to 9.5 mm. Antennæ 16 mm.

Fifth instar (Plate I, Fig. 5 and Plate III, Fig. b): Color pale yellowish green. Segments of abdomen with a fairly regular pattern of roundish white blotches; a small one on front and one on hind margin on median line; larger blotches on each side are arranged alternately near the front and hind margins. Outer side of hind femur with numerous black spots extending over the distal two-thirds or four-fifths. Antennæ marked similar to adult. Length 11 to 12 mm. Antennæ 23 mm.

Adult (Plate I, Fig. 6 and Plate V, Fig. a): Moderately slender. Pronotum as broad as long. Color very pale green. Top of head between eyes and antennæ

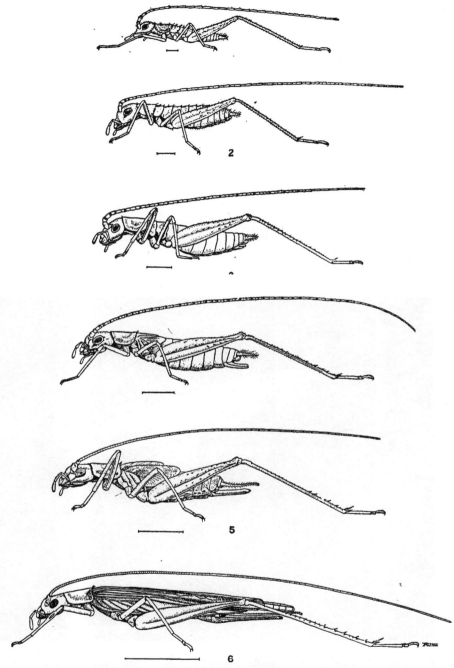

PLATE I.— LIFE STAGES OF *Œcanthus niveus*; NYMPHAL INSTARS AND ADULT.

PLATE II.— 1, FEMALE OF *Œcanthus niveus* FEEDING ON METANOTAL GLAND OF MALE;
2, CHARACTERISTIC POSTURE OF FEMALE WHEN OVIPOSITING.

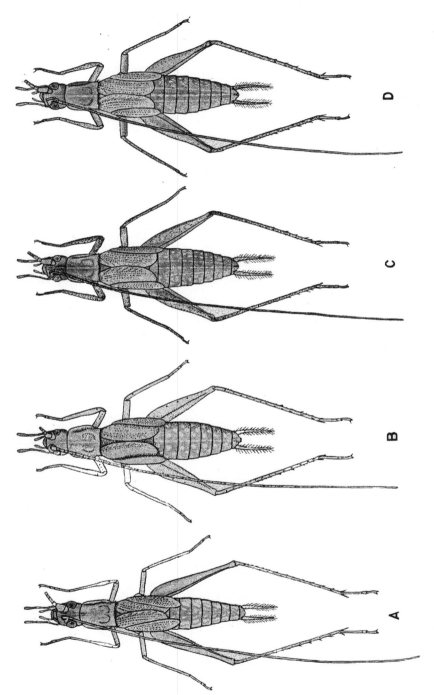

PLATE III.—FIFTH STAGE NYMPHS OF TREE CRICKETS: a, Œcanthus angustipennis; b, niveus; c, nigricornis; d, quadripunctatus.

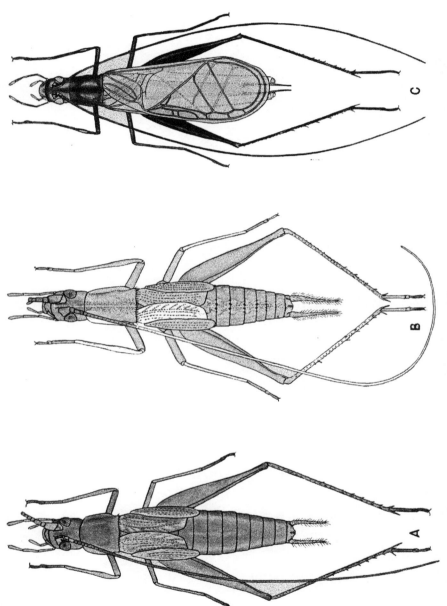

PLATE IV.—FIFTH STAGE NYMPHS OF TREE CRICKETS: a, *Œcanthus pini*; b, *latipennis*, c, ADULT MALE, *nigricornis*.

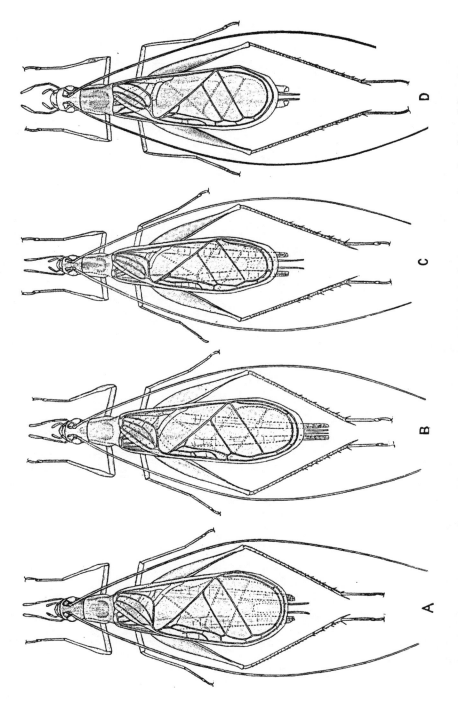

PLATE V.—ADULT MALE TREE CRICKETS: a, Œcanthus niveus; b, exclamationis; c, angustipennis; d, quadripunctatus.

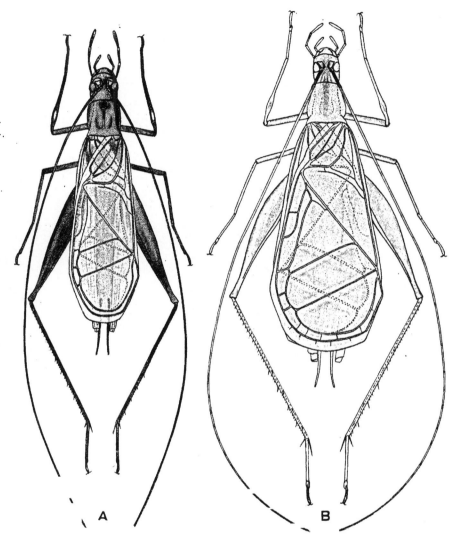

PLATE VI.—ADULT MALE TREE CRICKETS: a, *Œcanthus pini;* b, *latipennis.*

orange yellow; occipital area with longitudinal transparent greenish blotches separated by white lines. Wings transparent, with a slight greenish tinge; veins more or less colored with yellowish green. Fore wings of male very broad. Antennæ white, with gray annulations in the distal part at intervals of about four segments. First segment is pale orange yellow on all parts except the large swelling on the front and inner side which is white and has a conspicuous round black spot in the center. (Fig. 1, a.) The second segment is white with a similar spot. Length to end of abdomen 14 mm. Fore wing of male 13–14 mm. x 6 mm. Fore wing of female 12–13 mm.

SONG.

The song of *niveus* is one of the most conspicuous and musical of the insect sounds commonly noted in late summer and autumn. It can be heard from the time the insects commence to mature — early in August in this latitude — until they succumb to frosts of late October or early November. The song begins at the approach of darkness and continues until morning. Occasionally a few of the insects may be heard during the middle of day when the weather is very cloudy. The song consists of a monotonous series of clear, high-pitch trills rhythmically repeated for an indefinite length of time. The quality is that of a clear, mellow whistle and has best been described by the words, *treat — treat — treat.* The pitch varies somewhat with the temperature, but on an ordinary summer evening it is about C, two octaves above middle C, or on a warm evening it may reach as high as D. The rapidity of the notes is directly dependent on the temperature. On a very warm night we counted 155 beats per minute, while on a cool night the number was only 64. This phase of the subject has been studied by Bessey[21], Edes[22] and Shull[33] and formulas have been worked out by which the temperature can be computed from the average number of notes per minute.

The song of different individuals may vary also in quality, intensity, pitch and rapidity of notes. There is, however, a tendency for the insects in a restricted site — as a raspberry plantation, clump of bushes or a single tree or a small clump of trees — to sing in unison in one synchronous movement.

Scudder[15] has described *niveus* as having two distinct songs, one for day and one for night. This certainly is not true, for none of our species have more than one type of song. Scudder made observations on two different species and admits in the same article that at the time he did not distinguish between the species. His "day song" would apply to *nigricornis*, *quadripunctatus* or possibly *latipennis*, although the last does not usually sing by day. His "night song" describes well the song of *angustipennis*.

OVIPOSITION.

In ovipositing the female *niveus* prefers to stand on the upper side of a branch, and when on a vertical or sloping surface generally works with the head uppermost. Another peculiarity which has been observed in only one other species, namely *exclamationis*, is the use of excrement to plug up the puncture. Just before depositing the egg and while the ovipositor is imbedded for its full length in the bark, the female forces out a drop of excrement, which by stretching out the tip of the abdomen she fastens to the bark just below the hole. After withdrawing the ovipositor she moves back, picks up the drop with her mouth and places it over the opening. Several minutes are then spent in packing it in and smoothing it out (Fig. 14, c).

The eggs are deposited in the soft inner bark of a number of trees and bushes. In most cases a part of the egg lies in contact with the wood, which is grooved by the ovipositor but not drilled into to any extent. In the process of healing, the puncture becomes surrounded, in most plants, by a hard woody callus. In trees having a soft fleshy bark, *niveus* prefers to oviposit in branches from one to three inches in diameter. The eggs may be placed in almost any part of the bark, but the lenticels are favorite places, probably because they are less resistant to the drilling operations.

In bushes and in trees which have a hard bark covering the larger branches, the eggs are commonly laid in the smaller branches in thick places in the bark at the sides of buds or small twigs. In raspberry canes, where this species is sometimes fairly common, the eggs are deposited singly in the fleshy area at the side of the bud in the axil of the leaf, and hey never extend through the woody layer into the pith (Fig. 14, d).

In breeding-cage experiments from one to thirteen eggs were deposited in a night by a single individual. A number of the insects laid a few eggs every night during the whole period of oviposition, while others suspended operations on several nights. The largest number of eggs deposited by a single female was seventy-five, the smallest number twenty-four, and the average of eleven individuals was forty-nine.

About Geneva the eggs of *niveus* are found most abundant in apple, plum and cherry trees, and are locally common in walnut and raspberry. One small elm tree was observed to contain a large number of them, and in this case the eggs were deposited in the corky areas of the bark and did not extend into the cambium layer. A few eggs were found in peach, witch-hazel, chestnut, butternut, wild crabapple, hawthorn, red oak, maple and lilac. Oviposition probably occurs in many other plants which possess bark of desirable thickness and not too resistant to the drilling operations of the insect.

ŒCANTHUS ANGUSTIPENNIS Fitch.

HISTORICAL NOTES.

This name was first applied by Fitch[4] to a single male specimen which he described as a variety of *niveus*. The description is very brief; and the only characters mentioned are that the tegmina are narrower than *niveus* and the hind wings protrude like tails. The latter character strongly suggests *Neoxabea bipunctata*, the male of which was unknown to Fitch, although the description might apply to *Œ. quadripunctatus* or the species now understood by this name. Beutenmüller[17] states that "whether the species has been correctly determined or not can never be definitely ascertained, as Fitch's type of the species, as well as his other species of *Œcanthus*, have been destroyed. I would propose that the name *angustipennis* nevertheless be retained for the species so well known by this name."

DISTRIBUTION.

Our knowledge of the extent of distribution of this species in New York is very limited. It is common in the lake region of the western part of the State and on Long Island, and probably the insect ranges over about the same territory as *niveus*. It has been recorded in literature from other states as follows: Massachusetts (Faxon), Connecticut (Walden), Georgia, Florida, Texas (Allard), Indiana (Blatchley), Illinois (Forbes), Kentucky (Garman), Kansas and Texas (Tucker), Minnesota (Lugger). From specimens examined we can record it from the following states: New Jersey, North Carolina, Florida (Amer. Mus.), Virginia (Schoene), Ohio (Kostir). Of the states mentioned, Minnesota represents the most northern limits of distribution, while Texas appears as the most western area of its occurrence.

HABITAT.

Angustipennis is quite often found in company with *niveus* and is generally abundant in apple orchards. It is more strictly arboreal than the latter species, and seems to be confined to woody plants, either trees or large bushes. About Geneva it has never been taken on raspberry, grape or weeds of any kind. Among forest trees it is more common than *niveus*. Many specimens have been collected from scrub and bur oaks on Long Island and from alder in a swamp near Geneva. Davis[49] states that in Florida he found this species among golden rods and other low plants by the road side and that they also occurred among the small oaks and other trees.

DESCRIPTION OF LIFE STAGES.

Egg (Fig. 15, b).— The eggs are white and average a trifle smaller than those of *niveus*. The cap is narrower than in the latter species and varies greatly in length. Short specimens (Fig. 15, d) measure about .4 mm. in length and breadth, while the long ones (Fig. 15, c) reach .7 mm. in length, have a broad base and taper down to a rather narrow tip. The projections of the cap are short and thick, measuring about .011 mm. in breadth by .014 in length. (Fig. 15, e.) The end of the cap is broadly rounded and the base slightly constricted.

The average measurements of twenty specimens of eggs are as follows: length, 2.77 mm.; greatest width, .51 mm.; length of cap, .48 mm.; width of cap, .42 mm.

Nymph.— First instar: Color white. Markings of head and thorax as in follow_ ing stage. Antennæ entirely white; occasionally with a dark spot on the inner edge of the first segment. Hind femora with a few black spots near distal end; hind tibiæ with a conspicuous black space at distal end covering about one-sixth of entire length. Length 3 to 3.3 mm. Antennæ 8 mm.

Second instar: Color greenish white. Head with a short black line above and back of each eye, and with black specks at the base of minute bristles between eyes and antennæ. Thorax with a pair of dark lines near the median line. First segment of

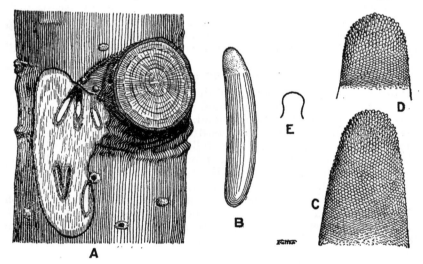

Fig. 15.— *Œcanthus angustipennis* Fitch.

a, Egg punctures in apple wood (x 3); b, egg (x 15); c, d, long and short egg caps (x 50); e, projection of egg cap (x 500).

antenna with a black spot on the inner edge. Distal half of antenna very faintly annulated. Hind femur with only four or five black spots on the outer side near the distal end. Length 4.5 to 5 mm.

Third instar: Dorsal area of abdomen pale green with a small median white spot on hind margin and a pair of white spots near front margin. Sides pure white. Basal antennal segment with the black spot on inner edge; and most specimens have a more or less distinct short line on the front side near the inner edge. Second segment with a small black spot on the front and inner side. Length 6 to 7 mm.

Fourth instar: Pale green. Head slightly yellowish above. Two median longitudinal lines of pronotum faint. Median area of abdominal segments pale yellowish green; the three white spots are relatively small. Upper part of side of each segment with a large elongate white spot reaching from front to hind margin, constricted or divided in the middle and surrounded by a ground color of pale yellow. Sides below are pure white. Antenna with a rather prominent white lump on the front and inner side and bounded on the outer side by a curved black mark. Second segment with an elongate spot. Length, 8 to 10 mm.

Fifth instar (Plate III, Fig. a): Top of head between eyes yellow or pale orange. Median area of pronotum greenish; with two faint dark median lines. Abdominal markings as in the fourth instar. White prominence on the first antennal segment, with a black J-shaped mark; and the second segment with an elongate spot. Hind femora with a few black spots near the extremity. Length 11 to 12 mm.

Adult (Plate V, Fig. c).— Very slender. Pronotum a little longer than greatest breadth. Color very pale green. Light specimens have the top of the head between the eyes and antennæ yellow, and have a faint gray longitudinal streak on the pronotum. Darker specimens have the top of head orange yellow or even burnt sienna and the streak on the pronotum is strong brownish gray. Wings transparent, with greenish tinge and greenish veins. Fore wings of male comparatively narrow. Antennæ faintly annulated with gray on the distal part. The first segment is yellowish with the exception of a white prominence on the front and inner side, which bears a black J-shaped mark, with the crook turned inward. (Fig. 1, b.) Length to end of abdomen 14 mm. Fore wing of male 10–12 mm. x 4–5 mm. Fore wing of female 12 mm.

SONG.

The song of this species is not so loud as that made by *niveus* and is of a more mournful quality. The pitch is about a half tone higher and ranges from C$\#$ to D$\#$; two octaves above middle C, depending on temperature and somewhat on individual variation. Like the song of *niveus* it is intermittent, but can be readily distinguished by its longer notes and rests and its non-rhythmical character. Each trill continues from one to five seconds, but most commonly it lasts for about two seconds. The periods of rest vary more and may be from one to eight seconds or longer. The total number of notes per minute varies, even from one minute to the next, and is generally not more than ten, but may occasionally run as high as fifteen. On one occasion a specimen alone in a cage was observed to trill continuously for a minute or more. Out of doors the song might be unnoticed by anyone not endeavoring to detect it. On trees where *angustipennis* occurs in equal abundance with *niveus*, its song is nearly drowned out by the synchronous beat of the latter species and only by listening intently can it be detected. So far as observed it sings throughout the night and remains silent during the day.

OVIPOSITION.

In trees where *niveus* deposits eggs in branches from one to three inches in diameter, the female *angustipennis* usually selects for this purpose a small branch of about a half or third of an inch in diameter. She generally drills into the thick wrinkled places in the bark where the small twigs branch. Although we have only collected the eggs from apple, it is probable that they occur in all plants on which the adults are known to exist. This species has not been observed to use a drop of excrement for plugging up the opening of the egg puncture. For this purpose she snips off little particles of bark here and there, chews them up and pushes them into the hole. Occasionally when the female has deposited an egg, she does not completely remove the ovipositor, but starts to drill again in a slightly different direction and places another egg near the first. When cut into, the two punctures appear in the form of a V (Fig. 15, a). After an examination of a large number of egg punctures in apple orchards near Geneva only a few paired eggs were found, and caged crickets from this section deposited very few eggs in this manner. Apple branches from West Virginia and Kentucky contained a large number of the double punctures as well as single ones, and live specimens from Kentucky deposited fully half their eggs in pairs. The crickets of the two localities appear to be identical in every respect and this difference in habit seems to be merely a physiological variation.

ŒCANTHUS EXCLAMATIONIS Davis.

HISTORICAL NOTES AND DISTRIBUTION.

This species was first described by Davis [32] from specimens collected at Staten Island and several localities in New Jersey. It has since been recorded by Engelhardt from Johnson City, Tenn., and by Walden from New Haven, Conn. Mr. C. L. Metcalf writes that the North Carolina Department of Agriculture has a record of its occurrence in the western part of the state. The author has examined specimens taken by H. H. Knight at Hollister, Missouri, and by W. J. Kostir at Cedar Point, Sandusky, Ohio, and has collected the species with Mr. Davis at Central Park, Long Island, where it is fairly common.

HABITAT.

The only locality in which the author has any knowledge of the natural habitat of the species is at the edge of an area of natural prairie land on Long Island, northwest of the village of Central Park. In the zone separating the pine barrens from the open prairie the trees are scattered, and the predominant species are black jack oak (*Quercus marilandica*) and bur oak (*Q. macrocarpa*). *Exclamationis* was found in much greater numbers on the latter oak and in company with *angustipennis*. It was not collected from any other plants in this locality although many other trees were beaten and the weeds and grass were swept with a net.

DESCRIPTION OF LIFE STAGES.

Egg (Fig. 16, d).— The color is dull white, semitransparent and often yellowish when first laid. The cap (Fig. 16, b) is white but sometimes stained reddish from the bark. It is generally a little broader than long and the sides are parallel at base only and the tip is not so broadly rounded as that of *niveus*. The cap bears no spicules but is covered with low rounded elevations which in aggregate appear like scales. Those projections are about .02 mm. in diameter and .005 mm. high (Fig. 16, c).

The average measurements of twenty eggs are as follows: length 2.85 mm.; greatest width, .56 mm.; length of cap, .46 mm.; width of cap, .49 mm.

No specimens were obtained earlier than August 12 and at that time all had matured so that the nymphal stages could not be described.

Adult.— (Plate V, fig. b)   Color nearly pure white, with a slight tinge of greenish yellow in places.   Top of head between eyes and antennæ pale to medium yellow. Wings with yellowish veins.   Antennæ pale, without gray annulations.   First segment of antenna with a white opaque prominence on the front and inner side, which bears

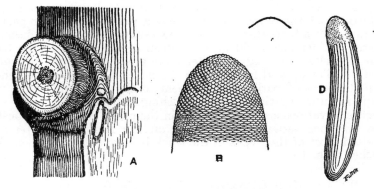

FIG. 1C.— *Œcanthus exclamationis* DAVIS.
a, Egg in oak (x 3); b, egg cap (x 50); c, projection of egg cap (x 500); d, egg (x 15).

a black club-shaped mark.   Second segment with a shorter black dash directly above; the two marks appear like an inverted exclamation point.   The hind pair of wings are generally long and equal the cerci.   Length to end of abdomen 14–15 mm.   Fore wing of male 13–14 mm. x 5–5.5 mm.   Fore wing of female 13–14 mm.

SONG.

The song of *exclamationis* is intermittent and non-rhythmical and most resembles the song of *angustipennis*.   The pitch is the lowest of any of the species studied, and reaches only to the second B above middle C.   The common length of note and rest is two or three seconds but this varies much, as in the song of *angustipennis*.   The beginning of each note is comparatively weak, but the sound increases in volume and slightly in pitch and continues uniformly until it abruptly ends.   In quality it most resembles the distant singing of the common toad.

It was observed that the males in the cages kept moving about while singing and that if no females were near they lowered their wings to the normal position between each note.   If a female was close by, the male became more excited, kept the wings elevated and repeated the notes in more rapid succession.

OVIPOSITION.

No bur oak (*Q. macrocarpa*) could be obtained near Geneva, so that the females of this tree cricket were not supplied with the wood in which they would naturally oviposit.   However they deposited many eggs in branches of red oak which were left in the cages. Whether the bark of the latter oak is more resistant than bur oak was not determined, but it seemed to require considerable labor for the females to bore the holes, and the whole process of egg laying generally continued for about an hour.. In some cases the cricket seemed to be unable to finish the hole and would pull out the ovipositor and chew a pit in some other part of the branch for a fresh start. In the few cases of oviposition observed in detail the female would fasten a drop of excrement to the bark before depositing the egg and after pulling out the ovipositor would pick the pellet up and force it into the hole. She would then chew off small pieces of bark and add them to the plug, and spend five or ten minutes putting on the finishing touches. Although oak branches of various sizes were placed in the cages, the females mostly chose branches between half an inch and an inch in diameter and deposited most of their eggs in the thick bark about the base of side branches (Fig. 16, a).

ŒCANTHUS QUADRIPUNCTATUS Beutenmüller.

HISTORICAL NOTES.

For many years this light colored species had been regarded as merely a pale variety of *nigricornis*: Beutenmüller[17] in 1894 named it as a new species, his claim being that of the many pairs of tree crickets he had observed mating in the field, in no case were these two forms together. Many entomologists today believe that they are only varieties, and there can be no doubt about their close relationship, but careful studies on the two forms have revealed constant differences in habits as well as in body characters, which would indicate that they are quite distinct species.

DISTRIBUTION.

Quadripunctatus is a common species in most parts of New York State, with the exception of the northern forested areas. Localities in North America have been recorded as follows: Massachusetts (Faxon), Connecticut (Walden), Long Island and New Jersey (Davis), Ontario (Walker), Indiana (Blatchley), Illinois (Forbes), Minnesota (Lugger), Manitoba (Walker), Maryland (Rehn), Georgia (Allard), Kansas and Texas (Tucker), Texas and Colorado (Caudell). From specimens examined it can be recorded from the following states: Massachusetts, Connecticut, New Jersey, North Carolina and Florida (Amer. Mus.), Ohio (Kostir), Minnesota (Patch), Colorado and Utah (Titus) and Arizona (Morrill).

HABITAT.

Although *quadripunctatus* can sometimes be found in the same locality with *nigricornis* the preferred habitat and places of maximum abundance of the two species are quite different in character. Upland fields abounding in medium sized weeds such as aster, sweet clover, daisy, golden rod, ragweed, and especially the wild carrot or Queen Ann's lace (*Daucus carota*), form the favorite environment of this species. Quite often in a field of this character where *quadripunctatus* is prevalent, the brushy fence rows surrounding the area will be inhabited by *nigricornis*.

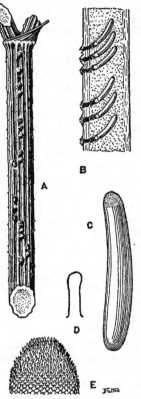

DESCRIPTION OF LIFE STAGES.

Egg (Fig. 17, c).— The color when the eggs are first laid is bright golden yellow but becomes somewhat paler as development goes on. In general they are a trifle yellower than the eggs of *nigricornis*. The eggs average smaller than those of the other species; they are about equal in length to those of *angustipennis* but are more slender. The cap is pure white and comparatively small (Fig. 17 e). In shape it is hemispherical but a little more prolonged than the cap of *nigricornis* eggs. The projections of the cap are long and slightly thickened at the end (Fig. 17 d). Those near the end of cap measure .02 to .025 mm. in length, by .007 to .009 mm. in diameter. Measurements of fifty specimens of eggs average as follows: length, 2.8 mm.; greatest width, .45 mm.; length of cap, .30 mm.; width of cap, .35 mm.

Nymph.— First instar: Color, pale slightly greenish yellow. A slight infuscation extends along the dorsal side from the antennæ back, and is divided along entire length by a narrow pale median line. Just back of the antennæ the median line meets a pale transverse curved line which arches posteriorly. On the abdomen the shaded area is bounded on each side by a pale line which is in turn bordered by a faint dark line. The antennæ are gray all over but darkest toward the extremity. Length 3 mm. Antennæ 6 to 6.8 mm.

Second instar: Pale greenish yellow with scattered whitish flakes. Pale dorsal median line present, dorsal infuscation very faint. The two basal antennal segments are pale in color and the first segment has a dark longitudinal streak on the inner edge of the front side. Length 4 to 5 mm.

Third instar: Greenish yellow, mottled with small whitish spots and with a pale median line. Markings of basal antennal segments of the same pattern as in adult but faint. Length 6 to 7 mm.

Fourth instar: Greenish yellow. Most of hairs on body pale in color. Tibial spurs black at tip only. Last two segments of hind tarsi nearly black. Length 8.5 to 10 mm.

Fifth instar (Plate III, Fig. d): Greenish yellow. Body hairs mostly pale. Brown spots at bases of hairs on legs, cerci, top of head and median area of pronotum. Markings on basal segments of antennæ more distinct. Length 11 to 12 mm.

Adult (Plate V, Fig. d).— Body pale yellowish green. Head more yellowish and faintly streaked with green. Abdomen yellowish above and below and greenish

Fig. 17.—*Œcanthus quadripunctatus* BEUTENMÜLLER.

a, Egg punctures in wild carrot (x 1½); b, longitudinal section of the same (x 3); c, egg (x 15); d, projection of egg cap (x 500); e, egg cap (x 50).

on sides. *Ventral area not darkened.* Legs dull yellowish green. Antennæ pale yellow at base, outer segments shaded with dull brownish gray which grows darker toward tip. The basal antennal segment is flattened transversely and along the inner edge of the front face is a narrow black line and at the distal end near the middle of the segment is a black spot. The second segment has a pair of black spots on the front side. Forewings of the male are transparent, with greenish yellow veins. Forewings of female with numerous green veins. Ovipositor brownish with black tip. Length to end of abdomen, 12–14 mm. Forewing of male, 11 by 4.5 mm. Forewing of female, 10–12 mm.

SONG.

The crickets of this species are more diurnal in their habits than those heretofore discussed, and they can often be heard singing at midday. Late in the afternoon more individuals join the chorus and they continue to sing throughout the night. The sound is a continuous, shrill whistle with a pitch about the third F # above middle C. It most resembles a small tin whistle but with a slight rasping or quavering quality. The call continues generally for a period of several minutes without ceasing, and in a field where large numbers are present it gives rise to a shrill, diffused ringing from which the individual sounds can not be distinguished. On cold nights the pitch drops a little and the sound becomes very faint. The call of old males often becomes a weak rasping shuffle. The continuous note distinguishes the song of this species and those following from that of the three preceding species, which have an intermittent call.

OVIPOSITION.

For the purpose of oviposition the female *quadripunctatus* selects small pithy weeds. She nearly always stands head downward on the stalk while depositing the eggs, which consequently are directed upward from the opening (Fig. 17, b). Out of a hundred eggs examined to determine this point only a short row of four were directed downward. The slant of the egg in the pith depends somewhat on the diameter of the stalk. The cap end lies within one-half to one millimeter from the opening and the inner end generally touches the wood on the opposite side if the stalk is not too wide. When the pith is of large diameter the eggs lie at about an angle of forty-five degrees with the long axis of the stalk, but if the pith is narrower the eggs lie more nearly parallel to the stalk, and in some cases the pith cavity is completely filled by the egg. The female works on the upper side of a leaning stem and deposits several eggs in a row. If the stalk happens to be vertical, as it seldom is, the eggs may be thrust in from different directions. The rows do not have the compactness and regularity to be observed in the oviposition of *nigricornis*, but are crooked and broken up into isolated punctures and short rows of from two to five punctures, each group separated by gaps from three to fifteen millimeters long (Fig. 17, a). Within the short rows the holes are placed about one to each millimeter, and seldom more than eight or ten are found in one group. In some cases the eggs are so scattered along the stem that they can scarcely be said to form a row. After the egg is deposited the female chews off a small patch of the outer tissue of the stalk and pushes the material into the hole at the same time pressing in the frayed edges. The neatly capped puncture is a very inconspicuous object. In the choice of plants for oviposition this species is very partial to the wild carrot (*Daucus carota*). So confined are they to this plant, at least in this region, that after many long searches only two other plants were found bearing their eggs. One of these was a small plant of golden rod and the other a purple aster. Doubtless other small plants with a solid pith are used in localities where the wild carrot is not so abundant. The size of stalk used varies from two to five millimeters in diameter and in this respect the insect shows a marked difference from the closely related *nigricornis*.

ŒCANTHUS NIGRICORNIS Walker.

HISTORICAL NOTES.

This insect is very generally known in literature under the name *Œ. fasciatus* (De Geer), Fitch, since it was first described by Fitch under that name. It has however been pointed out by Scudder[5] and later by Beutenmüller[17] that Fitch erroneously identified his insect as De Geer's[1] *Gryllus fasciatus* which is really a *Nemobius*. Therefore he did not give a name to his species and Beutenmüller recommended that the name *nigricornis* be used instead. The original description of the latter by Walker[8] in 1869, fits this insect very well and is undoubtedly taken from the same species. As stated before Walsh[6] and Riley[7, 10] considered this speci s as a dark variety of *niveus* and many collectors at the present time consider this tree cricket and *quadripunctatus* as varieties of the same species.

DISTRIBUTION.

This tree cricket is very common and is widely distributed over New York and throughout the United States. From literature it is recorded as follows: Massachusetts (Faxon), Connecticut (Walden), New Hampshire (Henshaw), New Jersey (Davis), Ontario (Walker), Tennessee (Morgan), Mississippi (Ashmead), Michigan (Allis), Indiana (Blatchley), Illinois (Forbes), Minnesota (Lugger), Nebraska (Bruner), Oklahoma and Arizona (Caudell), Texas and Kansas (Tucker), California (Baker). From specimens examined we can record it from the following states: New Jersey and Connecticut (Amer. Mus.), North Carolina (C. L. Metcalf), Ohio (Kostir).

HABITAT.

In general *nigricornis* prefers larger plants and more dense growth than *quadripunctatus*, and its habitat is more varied. While the latter is confined largely to upland weedy fields, the former can be found in tall, rank growths of weeds such as in swamps and river bottoms, in brush or second growth, in fence rows grown up with bushes, briars and vines, in raspberry plantings, vineyards, nurseries and even occasionally in orchards. While the list of plants used for oviposition, given in a following paragraph, contains many names of trees, it has been our experience that these records are of rare or local occurrence and generally refer to small seedling trees growing among brush, and thus do not convey the proper impression regarding the habitat of the insect.

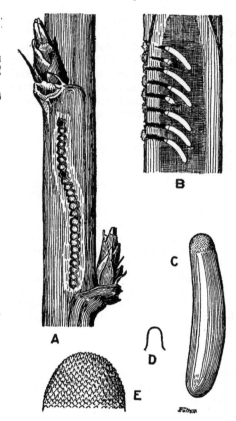

FIG. 18.— *Œcanthus nigricornis* WALKER.

a, Egg punctures in raspberry (x 1½); b, longitudinal section of the same (x 3); c, egg (x 15); d, projection of egg cap (x 500); e, egg cap (x 50).

DESCRIPTION OF LIFE STAGES.

Egg.— The eggs (Fig. 18, c) are of a light or medium yellow color, and brightest when first laid. The cap is rather small, broader than long and hemispherical, the sides being parallel only at the extreme base. (Fig. 18 e.) The color of the cap is dull white but is sometimes stained reddish when in certain plants. The projections are short, cylindrical and rounded at the tips. (Fig. 18, d.) Those near the end of the cap are .012 mm. long by .008 mm. in diameter. The eggs of this species can be distinguished from those of *quadripunctatus* by the much shorter projections on the cap. Average measurements of thirty-six specimens are as follows: Length 2.9 mm.; greatest width .57 mm.; length of cap .33 mm.; width of cap .44 mm.

Nymph.— First and second instars: In these two stages the crickets seem to be identical in every respect with those of *quadripunctatus*.

Third and fourth instars: The differences between *nigricornis* and *quadripunctatus* in these two stages are very slight and only relative. The former is more nearly pure green with less yellowish tinge; dorsal infuscation is usually plainly visible; hind legs, especially the tibiæ, more heavily speckled with black dots; and the black line on the basal antennal segment is broader. These statements are based on averages only. Not knowing where any given specimen was found it would be impossible to state with certainty which species it belonged to.

Fifth instar (Plate III, Fig. c): This is the only nymphal stage in which the two species can be separated. *Nigricornis* averages a little larger, and is not so yellowish in color. Dorsal part of head is slightly brownish. Hairs of body are mostly black or brown instead of pale. The legs appear dark, due to dark hairs and spots. Spurs are black nearly to base. Antennæ are black or nearly so in outer part; the four spots at the base are large and conspicuous. The most constant character, which can not be relied on in dried specimens, is the dark band on the ventral side of the abdomen; the sternum of each segment is slightly infuscated and covered with small brownish spots.

Adult (Plate IV, Fig. c). — The amount of color in this species varies considerably and newly moulted specimens are lighter than old ones. The light specimens are greenish yellow. Head with blackish or sepia shading on median area, sides and front below the antennæ. Pronotum with similar shading on sides and median area. Wings clear with greenish yellow veins and tinge of green between veins on inner edge. Femora dull green; tibiæ and tarsi black. Antennæ black; first and second segments greenish yellow. The first segment has a brownish shading covering the inner and upper part of the front side, and including a heavy black line along the inner edge and a black spot near the distal end, which may be confluent with the black line. Second segment with two elongate black spots. (Fig. 1, f.) *Venter of abdomen solid black;* the remainder greenish yellow. Dark specimens late in the season have the head, pronotum, legs and antennæ nearly entirely black. Both pairs of spots on the two basal antennal segments are confluent (Fig. 1, g) and in some specimens both of these segments are almost entirely black. Length of body, 14 mm. Fore wing of male, 10-11 mm. by 4.5 to 5 mm. Fore wing of female 11-12 mm.

SONG.

The song of *nigricornis* so closely resembles that of *quadripunctatus* that it is very difficult to distinguish the two. On the average the song of the former is louder and shriller and with less rasping quality. The pitch of the two is the same, the third F # above middle C, and both species have the habit of singing during the day as well as at night. After a study of the habitats of the two species, however, one can judge fairly accurately in the field, which species he hears from the character of the surrounding vegetation.

OVIPOSITION.

In preparing for oviposition the female generally selects a position well above the ground. She works head uppermost, and if the stalk of the plant leans a little she prefers to operate from upper side. As with other tree crickets the first step in the process is to chew a pit in the bark to start the ovipositor. The length of the drilling operation depends on whether the eggs are being deposited in the stalk of an herbaceous weed or in the hard woody twig of a tree or bush and varies from ten minutes to half an hour. After the egg has been deposited and the mucilaginous substance discharged into the hole, the female removes the ovipositor and chews off pieces of the bark just above the puncture to plug up the opening. The pit made in so doing serves as a starting point for the next drilling operation. This is repeated a number of times and results in a long compact row of eggs, with from seven to ten punctures to each centimeter. (Fig. 18, a.)

The eggs slant across the pith cavity at an angle of from forty-five to sixty degrees with the long axis of the stalk. (Fig. 18, b.)

If the pith cavity is small and the woody layer thick, a part of the capped end of the egg is partly imbedded in the wood. When the punctures are close together the eggs are directed alternately to the right and left so that they do not interfere with each other. In contrast to the oviposition of *quadripunctatus* the eggs of this species are generally directed downward from the opening as the result of the female working head uppermost. In localities where strong prevailing winds have caused all the weeds to lean in one direction, the habit of the female in ovipositing on the upper side of the stalk might give rise to the impression that she always worked on the same side with reference to the points of the compass.

The oviposition period commences during the latter part of August and generally extends through the month of September. The total number of eggs deposited varies greatly with individual crickets. In 1910 the records of six pairs confined in breeding cages were respectively as follows: (1) 165 eggs, (2) 64, (3) 26, (4) 78, (5) 52, (6) 31. During 1913 three pairs deposited respectively 22, 51 and 60 eggs. The eggs were deposited in rows of from seven to twenty-one punctures. Occasionally the number of eggs in a series was increased over night or over a succession of nights at varying intervals by ovipositions by the same female. Observations in a patch of raspberries showed that the number of eggs in a row ranged from two to eighty-seven. The average number in nineteen rows taken at random was about thirty-two eggs.

This species prefers for the reception of its eggs plants which have a central pith surrounded by a woody outer layer, and there are a great many plants of this character which are selected by the insect for this purpose. Eggs are deposited most abundantly in raspberry, blackberry, *Erigeron canadensis* and the larger species of *Solidago*. They are also common locally in elder, grape, sumac and willow. A few eggs may occasionally be found in the twigs of peach [1] apple,[2] elm, maple and hickory. Mr. Goodwin of the Ohio Station writes that considerable oviposition by this species occurs in peach orchards and vineyards in northern Ohio, especially on trees and vines which adjoin uncultivated fields. Similar conditions with respect to vineyards have been noted in the grape-growing region in Chautauqua county, New York. Mr. W. T. Davis of Staten Island reports that he has also found eggs of this insect in wild cherry, white ash and *Baptisia tinctoria*. In going over the literature of this species we have found numerous descriptions of the work of this insect in various plants besides those given above, but always under the name of *niveus*. When the eggs are described as deposited in long rows there is little doubt as to their identity; for the only other widely distributed species with this habit is Œ. *quadripunctatus*,

---

[1] From material collected by J. L. King at Gypsum, Ohio.
[2] From material collected by B. G. Pratt, New York City.

which deposits eggs only in smaller and more delicate plants. On this assumption additional host plants as recorded in literature are currant, Helianthus, artichoke, Ambrosia, plum, cottonwood, box elder, cherry, dogwood, black locust, honey locust, sycamore and catalpa.

In the size of the stalk selected for oviposition *nigricornis* shows a distinct difference in habit from *quadripunctatus*, in that it almost always seeks one of larger diameter. With grape vines and certain weeds, stems less than five millimeters in diameter are seldom chosen and with raspberry canes and elder the common thickness of the wood is not much under a centimeter.

ŒCANTHUS PINI Beutenmüller.

HISTORICAL NOTES AND DISTRIBUTION.

This tree cricket was first described by Beutenmüller[16] from specimens collected in Windham Co., Conn. It has since been recorded as follows: Gloucester, Mass. (Henshaw); Riverton, N. J. (Rehn); Karner, N. Y. (Felt); Staten Island and Long Island, N. Y., and Bloomsburg, Columbia Co., Pa. (Davis).

HABITAT.

All the above records are from specimens collected on pine trees and apparently this species does not inhabit any other kind of plant. The author in company with Mr. Wm. F. Davis collected a large number of specimens near Central Park, Long Island. In this region as also in the New Jersey pine barrens the pitch pine (*Pinus rigida*) is the predominant species. The trees are rather scattered and closely surrounded by scrub oak and other smaller plants, but in no case were specimens taken on anything but the pines. Mr. Davis informs me that his specimens collected at Bloomsburg, Pa., were taken from the Jersey or scrub pine (*P. virginiana*).

DESCRIPTION OF LIFE STAGES.

Egg.— The eggs (Fig. 19, d) closely resemble those of *nigricornis*. The color is yellow but not so bright as the eggs of *quadripunctatus*. The cap (Fig. 19, e) is a trifle larger than that of *nigricornis* and the projections (Fig. 19, c) are a little larger and flattened, except those on the extreme tip which appear to be cylindrical. The shape of the projections is the most prominent difference between the eggs of the two species. This is most easily seen when the cap end of the egg is raised so that one gets an end view of the projections on the side. Those of *pini* appear transversely elliptical, with one diameter about double the other, while those of *nigricornis* appear round or broadly oval. The projections of the eggs of *pini* measure .018 mm. long, .012 mm. wide and .006 mm. thick. The average measurements of twenty eggs are as follows: Length 3.14 mm.; greatest width .55 mm.; length of cap .35 mm.; width of cap at base .47 mm.

Nymph.— Fifth instar: (Plate VI, fig. a.) Specimens were not obtained of any stage earlier than the fifth nymphal instar which is as follows: Pronotum short and broad. Head and thorax light brown, with obscure pale median line and somewhat paler on edge of disk of pronotum. Wing pads green, with dark green veins. Abdomen with a broad dorsal area of dull green, bounded by a streak of pale yellowish green on each side. Sides of abdomen dull green, fading out below. Sides and under part of thorax slightly pinkish. Antennæ yellowish on first two segments, remainder brown, growing darker toward tip; basal segments with black markings resembling *quadripunctatus*. Legs greenish brown; spines on hind tibiæ black. Length 14.5 mm.

Adult.— (Plate VI, fig. a.) Head and pronotum nearly uniform dull reddish brown; pronotum with a paler stripe on each side. Underside of thorax pale yellow with brown blotches. Abdomen with ventral area uniform dull brown, bordered on each side by a narrow, sharply defined cream colored stripe; sides dull brown; dorsal area greenish brown. Antennæ dull brown; two basal segments pale brown with black markings on the front side; first with a black line along inner edge and a transverse, slightly oblique line near the end; second with two elongate parallel black spots. (Fig. 1, e.) Legs dull, olive brown. Wings of male transparent, with green veins; area anterior to the rasp clouded with green; a stripe on the inner edge, beyond the rasp and one along the fold of the wing, solid yellowish green, so that when wing is raised it is seen to be clear, with two green stripes. Wings of female with bright green veins except along the fold where they are pale yellow, giving the effect of a cream colored stripe. Near the base they are tinged with reddish brown. Ovipositor dark brown with black tip. Length of body 15–16 mm. Fore wing of male 13 mm. by 5 mm. Fore wing of female 12 mm.

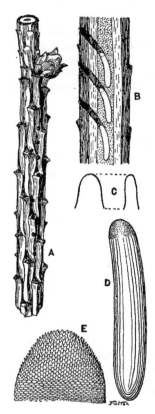

FIG. 19.— *Œcanthus pini* BEUTENMÜLLER.

a, Egg punctures in pine (x 1½); b, longitudinal section of the same (x 3); c, projection of egg cap, two views of the same structure (x 500); d, egg (x 15); e, egg cap (x 50).

SONG.

The song of *pini* is a shrill, continuous whistle, which closely resembles the call of *nigricornis* and *quadripunctatus*, but can be distinguished by its lower pitch. It is on the average about a note and a half lower than the song of those species, or the third E above middle C. Beutenmüller[18] describes the sound when many individuals are singing together, as " not unlike the jingling of sleigh-bells at a distance." The crickets in the cages in the laboratory generally began singing late in the afternoon and continued throughout the night.

OVIPOSITION.

The females in our breeding cages oviposited mostly at night but on a few occasions they were observed drilling in the daytime, both in the morning and afternoon. Before beginning to drill a pit is chewed in the bark as with the other species. The pine twigs are rather resistant and the drilling operation requires thirty or forty minutes. On completion of the process the hole is plugged up with bits of chewed bark. One female under observation deposited fourteen eggs in a single night; another, which was caged alone with a male, laid only thirty-one eggs in all. The eggs are placed in rows like those of *nigricornis* but are not so close together; one puncture to every three millimeters is a common distance. A favorite point for oviposition is at the end of one of the elongate scales of bark with which the pine twigs are covered. (Fig. 19, a.) The eggs may slant in either direction in the twigs but of the limited number deposited in the laboratory most of them were directed downward indicating that the females preferred to work head uppermost.

In the breeding cages the crickets were first supplied with white pine and Austrian pine, but later some branches of the pitch pine were procured. Some apple twigs and wild carrot stalks were also placed in the cages. The crickets showed a decided preference for pitch pine while it was in fresh condition but later oviposited in the wild carrot stalks and apple twigs. This was probably because the pine twigs became dry and hard and a fresh supply could not be readily obtained. In no case did the crickets oviposit in the other two species of pine. The size of the pine twigs used varied from three millimeters to nearly a centimeter in diameter. Eggs in the narrow twigs were found to lie in the small pith cavity while those in the larger branches were partly or wholly imbedded in the wood.

ŒCANTHUS LATIPENNIS Riley.

HISTORICAL NOTES AND DISTRIBUTION.

Riley's[10] original description of this insect is accompanied by a brief but correct account of the song and egg laying habits. He had described the eggs in grape in an earlier report[9] but at that time supposed they were the eggs of *Orocharis saltator*. His specimens were mostly from Missouri, but he had one from Alabama and one from South Texas. Other records obtained from literature, correspondence and collections are as follows: Long Island and New Jersey (Davis); Michigan (Allis); Ohio (author's collection); Indiana (Blatchley); Illinois (Forbes); Minnesota (Dugger); Nebraska (Bruner); North Carolina (C. L. Metcalf); Kentucky (Schoene); Tennessee (Morgan); Georgia (Allard). The list given here would indicate that the range of this cricket does not extend as far north as that of the other widely distributed species. As far as we know it has been collected in this state only in the southeastern part.

The author is more familiar with the species in southern Ohio, where eggs were obtained for life history studies. There the insect is common in weedy places, in flower beds and shrubberies about farm houses, and especially among grape vines. W. T. Davis writes that in Long Island, Staten Island and New Jersey he has found *latipennis* fairly common among the oak scrub.

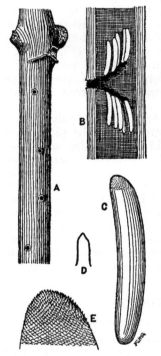

FIG. 20.— *Œcanthus latipennis* RILEY.

a, Egg punctures in g r a p e (x 1½); b, longitudinal section showing eggs in golden rod (x 3); c, egg (x 15); d, projection of egg cap (x 500); e, egg cap (x 50).

DESCRIPTION OF LIFE STAGES.

Egg.— The eggs (Fig. 20, c) are rather long and slender, and of a pale yellow color. The cap (Fig 20 e) is white but rather inconspicuous; it is broader than long and most resembles the cap of *nigricornis* but is generally a little oblique, that is, the extreme tip is nearer the side of the egg having the lesser curvature. The projections of the cap (Fig. 20, d) are narrow and a trifle longer than those of *nigricornis* and many are obtusely pointed at the tip rather than rounded. The long ones near the end of the cap measure .017 mm. long by .007 mm. in diameter. Measurements of thirteen specimens of eggs taken from several sets show considerable uniformity in shape and size. The averages are as follows: length 3.00 mm.; greatest width .50 mm.; length of cap .28 mm.; width of cap .40 mm.

Nymph.— First instar: Ground color of body pure white. A broad band of purplish red extends full length along the dorsal side and encloses a narrow white median line, which joins a transverse curved line between the eyes. On the concave anterior side of the latter white line is a small patch of red between the antennæ. Eyes are faintly yellow. Antennæ are unmarked. White median line on the abdomen is bordered on each side by a dark brown line of about the same width. Hind margin of each segment with a pair of conspicuous black bristles on the brown lines; all other hairs on abdomen pale and scarcely visible. Cerci with a few specks of red. Legs unmarked; last segment of hind tarsi dark. Length 3.5 mm.

Second instar: Ground color of body slightly yellowish, most pronounced along the sides. Reddish band consists of a pink ground color much mottled and speckled with red, darkest along the median white line and edges. White line widened into a spot on hind border of each abdominal segment. Reddish brown line on each side less well defined than in first instar. Hairs and bristles on red band dark. Cerci pinkish at base. Antennæ white; segments in outer half very indistinct. Cerci pinkish at base. Last segment of hind tarsi black. Length 5 mm.

Third instar: Ground color pale yellow, strongest along sides. Median dorsal band of pale pink ground color, thickly speckled with reddish spots which occur at the bases of minute black bristles. Top of head posterior to and between the eyes marked with longitudinal reddish streaks. In thorax the pink blends into yellow on the sides but on the abdomen the pink area is outlined by a reddish line which is scalloped or curved toward the median line on the hind part of each segment. Narrow median pale line present, but obscure in places. Posterior edge of each abdominal segment with a white elevated spot on the median line, bounded on each side by smaller black spots, bearing tufts of black bristles. Antennæ pink at base, fading out into white on segments 4 to 6. Length 7 mm.

Fourth instar: General color pale greenish yellow. Outer ends of the transverse pale line between eyes, curl strongly forward. Median reddish band obscure on thorax. Abdomen yellowish green on sides with whitish blotches; upper part becomes uniform pale yellow and lower part cloudy white. Pinkish median band much modified on abdomen. Each segment bears a vase-shaped figure, broad and rounded on the anterior part, constricted just back of the middle and flaring out again on the hind margin. Ground color of figure pale pink, except in the enlarged part where it blends into a spot of pale greenish yellow in the center. Figure has a reddish margin and bears a number of minute dark bristles with a reddish spot at the base of each. Median line paler and slightly elevated; near the hind margin of each segment it is produced into a small white elevation at each of which is a bristly black spot. Length 10.5 mm.

Fifth instar: (Plate IV, fig. b). Head light yellow; upper part from antennæ back dull pink; transverse and median pale lines as in earlier instars. Pronotum yellowish, strongest on sides; dull median band made up of small black hairs and brownish specks. Abdomen colored similar to preceding stage; ground color of vase-shape figures on segments very pale or even white. Basal segment of antennæ yellowish mottled with pink; following segments reddish but fading to white within four or five millimeters. Legs greenish, unmarked. Last segment of hind tarsi black. Length 12 to 13 mm.

Adult (Plate VI, fig. b).— Body whitish. Head and pronotum tinged with yellow. Top of head from antennæ back with a pink patch, occipital region with four longitudinal dusky streaks fading out in front. Pronotum with a median dusky band, darkest in front; this is plain in some specimens and absent in others. Antennæ red at base, fading out near the eighth segment; the first segment has pale yellow ground color mottled with red especially along inner edge. Remainder of antennæ white faintly annulated at intervals with gray; extreme distal part dark gray. Legs white; knees yellowish; hind femur with a small black spot just before distal end. Wings of male transparent; in some specimens the large cells near the tip are bordered by slightly infuscated lines running near and parallel to the veins. Length of body 15–17 mm. Fore wing of male 13 to 16 mm. by 6.5 to 8. mm. Fore wing of female 13–14 mm.

SONG.

The song of *latipennis* is louder than that of any other tree cricket discussed in this bulletin. It is a clear, continuous whistle resembling the call of the preceding species, but is of a lower pitch, and has a more musical and bell-like quality. On an ordinary summer evening the tone is about the third D$\sharp$ above middle C, but on cool nights the pitch may drop a half note lower. In southern Ohio where the author has observed this cricket in the field, the males began singing in full chorus at dusk and continued throughout the night. The reared specimens in breeding cages at Geneva were much more shy than the other species and would not begin to sing until dark and when all was quiet about the laboratory.

OVIPOSITION.

This insect deposits its eggs in a peculiar manner which seems to possess some advantages over the methods employed by other tree crickets. In depositing the first egg in any part of the stalk the process is essentially the same as that of the three preceding species except that the female seems to prefer working on the underside of a stalk and may start with head either up or down. After the egg is placed the female turns around and chews at the hole for several minutes but does not plug it up. She then inserts the ovipositor in the hole and begins to drill, this time in the opposite direction. Not having the hard woody layer to penetrate, this second drilling reams out a hole in the pith in about five minutes and a second egg is deposited. The insect then turns around, chews at the hole and starts to drill again in the original direction. This process is continued, the female turning around after each drilling until several eggs have been deposited on each side of the hole. (Fig. 20, b.) Apparently no attempt is made to plug up the opening, which, on account of the repeated drillings, is large and conspicuous.

The eggs of this cricket were collected in southern Ohio in grape vines and in large stalks of golden-rod growing near the vines. The parts of the plants most used were the young shoots of the vines, from three to five millimeters thick and the upper ends of the golden-rod stalks where they were about five millimeters in diameter. The holes were most often on the under side and in many cases were arranged in rows, with intervals of about one centimeter or a little more. (Fig. 20, a.) The eggs of each of the two groups lie side by side in the pith and about parallel to the long axis of the vine or stalk. The size of the pith determines the number of eggs that can be placed together. The grape vine has a small pith and will hold only two or three eggs on a side, while in the golden-rod as many as six can be found in a single cluster. Thus a single hole in the outer woody layer may serve for the deposition of from four to twelve eggs.

NEOXABEA BIPUNCTATA De Geer.

HISTORICAL NOTES.

It seems desirable to include in this bulletin an account of the single described species of *Neoxabea,* since it is so closely related to the crickets of the genus *Œecanthus* that it was for a long time considered as a species of the same genus. Riley[10] regarded the differences between this species and the other tree crickets of generic importance and included it in the same genus with *Xabea decora* Walker from Sumatra, to which it seemed more closely allied. Later Kirby[31] established it in a genus by itself under the name of *Neoxabea.*

DISTRIBUTION.

Judging from the meager records of the species in this state it appears to be confined to the southeastern corner. The collection of the American Museum contains specimens from the Ramapo

Mountains, Rockland County, and from Southampton, Long Island. Other records obtained from literature and collections are as follows: Connecticut (Walden); New Jersey (Smith); Pennsylvania (De Geer's type), Ohio (King), Indiana (Blatchley), Illinois (Forbes), Kansas (Tucker), North Carolina (Amer. Mus.), Georgia (Allard), Nicaragua (Baker). Apparently these records indicate that most of New York is too far north to be within the range of this insect.

DESCRIPTION OF LIFE STAGES.

Egg.—Very long and slender and of a pale, transparent yellowish color. The cap is indistinct and consists only of an opaque, minutely roughened portion of the egg, and bears no projections as is common in *Œcanthus* eggs. Near the base the roughness takes the form of shallow depressions arranged in spiral rows, but toward the tip the rows break up and the surface is rugulose. The eggs measure from 3.85 to 4.10 mm. in length and from .45 to .50 mm. in diameter.

Adult.— The generic characters are as follows: hind tibiæ without teeth or spurs except at tip; basal segment of antennæ with a blunt tooth on anterior side; maxillary palpi with third and fifth segments very long and fourth short; hind wings very long and protrude behind body like tails. The general color is pale pinkish brown. The fore wings of the female bear two large dark brown blotches, one in front of the other. Male wings unmarked. Legs pale pinkish color. Length of body about 16 mm. Fore wings 13 mm. Hind wings 20 mm.

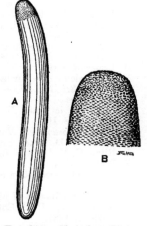

Fig. 21.— *Neoxabea bipunctata* De Geer.

a, egg (x 15); b, egg cap (x 50).

HABITS.

Our knowledge of this insect is confined to observations on a few adult females collected at Cedar Point, Ohio, by J. L. King of the Ohio Experiment Station. The lack of males precluded any observations on the song of the insects but did not prevent the females from depositing eggs. Not knowing at the time the natural host plant of the species it was necessary to place in the cages a number of different kinds of plants, among which was an apple branch about half an inch in diameter. This was selected by the females for oviposition and the eggs were deposited in the bark in about the same manner as those of *niveus*, except that most of them were directed upward from the hole. The crickets seemed to adapt themselves quite readily to the apple, but at the place where they were collected no apple trees exist, and I have since been informed by the collector that the insects were found on oak, willow and wild grape vines at a height of five or ten feet. The trees were along the forest border or standing isolated at the edge of open, sandy areas.

ACKNOWLEDGMENTS.

The author wishes to express his obligations to numerous entomologists for cooperation in determining the distribution of the insects.   He is greatly indebted to Mr. W. T. Davis of Staten Island for information on the habits of the crickets and assistance in collecting on Long Island; to Mrs. L. L. Van Slyke of Geneva for notes on the musical qualities; and to Mr. J. L. King of the Ohio Experiment Station for furnishing living specimens of *Neoxabea bipunctata*.

BIBLIOGRAPHY.

This includes the literature referred to in the bulletin and the more important articles on the habits of tree crickets.   All references to literature in the text bear the numbers listed below.

1. De Geer, Charles
     Memoir pour servir a l'histoire des insectes, T. III, pp. 522–523, 1773.
2. Lespès, Charles
     Mémoire sur les spermatophores des grillons,
     Ann. Sci. Nat., 3: 366–377, 1855.
3. Lespès, Charles
     Deuxieme note sur les spermatophores du *Gryllus sylvestris*,
     Ann. Sci. Nat., 4: ——, 1855.
4. Fitch, Asa
     3d Rpt. in Trans. N. Y. Agr. Soc., pp. 404–416, 1856.
5. Scudder, S. H.
     Jour. Boston Soc. Nat. Hist., 7: 430–432, 1862.
6. Walsh, B. D. *Pract. Ent.*, 2: 54, 1867.
7. Riley, C. V. 1st Rpt. Insects of Mo., p. 138, 1869.
8. Walker, Francis
     Cat. of Dermaptera Saltatoria of Brit. Mus., I.   1869.
9. Riley, C. V.
     5th Rpt. Insects of Mo., p. 120, 1873.
10. Riley, C. V.   U. S. Ent. Com., Bul. 6, Sup. to 9 Rpts. on Insects of Mo., pp. 60–61, 1881.
11. Riley, C. V.   Standard Nat. Hist., II, 1884.
12. Murtfeldt, Mary E.   *Insect Life* 2: 130–132, 1889.
13. Bruner, L.   Ann. Rpt. Nebr. State Bd. Agr., pp. 227–230, 1889.
14. Hart, C. A.   *Ent. News*, 3: 33, 1892.
15. Scudder, S. H.   23rd Rpt. Ent. Soc. Ont., f. 1892, pp. 65–67, 1893.
16. Beutenmüller, W.   *Jour. N. Y. Ent. Soc.*, 2: 56, 1894.
17. Beutenmüller, W.   Amer. Mus. Nat. Hist., Vol. 6, Art. 11: pp. 250–251, 1894.
18. Beutenmüller, W.   Amer. Mus. Nat. Hist. Vol. 6, Art. 12: pp. 269–272, 1894.
19. Davis, W. T.   *Jour. N. Y. Ent. Soc.*, 3: 142, 1895.
20. Lugger, O.   Minn. Agr. Exp. Sta., Bul. 55, pp. 358–363, 1897.
21. Bessey, C. A. and E. A.   *Amer. Nat.*, 32: 263–264, 1898.
22. Edes, R. T.   *Amer. Nat.*, 33: 935–938, 1899.
23. Faxon, W.   *Psyche*, 9: 183, 1901.
24. Blatchley, W. S.   Orthoptera of Indiana, 27th Ann. Rpt. State Geol. Ind., f. 1902, pp. 443–453, 1903.
25. Houghton, C. O.   15th Rpt. Del. Agr. Exp. Sta., pp. 150–152, 1903.
26. Garman, H.   Ky. Agr. Exp. Sta., Bul. 116, pp. 63–67, 1904.
27. Walker, E. M.   *Canad. Ent.*, 36: 253–255, 1904.
28. Forbes, S. A.   23rd Rpt. State Ent. Ill. pp. 215–222, 1905.
29. Hancock, J. L.   *Amer. Nat.*, 39: 1–11, 1905.
30. Felt, E. P.   N. Y. State Mus. Mem., 8: 602 and 698–700, 1906.
31. Kirby, W. F.   A Synonymic Cat. of Orthoptera, II: 72–76, 1906.
32. Davis, W. T.   *Canad. Ent.*, 39: 173, 1907.

33. Shull, A. F.   *Canad. Ent.*, **39**: 213–225, 1907.
34. Houghton, C. O.   *Ent. News.*, **20**: 274–279, 1909.
35. Houghton, C. O.   *Canad. Ent.*, **41**: 113–115, 1909.
36. Jensen, J. P.   *Canad. Ent.*, **41**: 25–27, 1909.
37. Jensen, J. P.   *Ent. News.*, **20**: 25–28, 1909.
38. Parrott, P. J.   *Jour. Econ. Ent.*, **2**: 124–127, 1909.
39. Allard, H. A.   *Proc. Ent. Soc. Wash.*, **12**: 36–38, 1910.
40. Baumgartner, W. J.   Kans. Univ. Sci. Bul., **19**: 323–345, 1910.
41. Cholodkovsky, N.   Ueber die Spermatophoren, besonders bei den Insecten, *Trav. Soc. Nat. St. Petersburg*, **41**: 72–78, Rés. 128–129, 1910.
42. Hancock, J. L.   Nature Sketches in Temperate America, pp. 379–386, 1911.
43. Jensen, J. P.   *Ann. Ent. Soc. Amer.*, **4**: 63–66, 1911.
44. Parrott, P. J.   *Jour. Econ. Ent.*, **4**: 216–218, 1911.
45. Allard, H. A.   *Ent. News.*, **23**: 460–462, 1912.
46. Boldyrev, B. Th.   Ueber die Bagattung und die Spermatophoren bei Locustodea und Gryllodea, *Rev. Russe d'Ent.*, **13**: 484–490, 1913.
47. Parrott, P. J., and Fulton, B. B.   *Jour. Econ. Ent.*, **6**: 177–180, 1913.
48. Engelhardt, Dr. V. v.   Ueber die Hancocksche Druese von *Œcanthus pellucens* Scop.   *Zool. Anz.*, **44**, **5**: 219–227, 1914.
49. Davis, Wm. T.   *Jour. N. Y. Ent. Soc.*, **22**: 204, 1914.
50. Parrott, P. J.; and Fulton, B. B.   N. Y. Agr. Exp. Sta., Bul. 388, 1914.

TECHNICAL BULLETIN No. 43. MAY, 1915.

New York Agricultural Experiment Station.

GENEVA, N. Y.

HUMAN MILK.

ALFRED W. BOSWORTH.

PUBLISHED BY THE DEPARTMENT OF AGRICULTURE.

HUMAN MILK. *

ALFRED W. BOSWORTH.

INTRODUCTION.

The knowledge of the chemistry of human milk has been derived from a general study of the constituents present and may be summarized by a statement of the amounts of fat, protein, sugar, and salts present. As a result of our somewhat detailed study of cow's milk, methods have been elaborated which allow of a more specific study of the milk constituents and enable us to form more definite opinions as to the condition in which they are present in milk. As these methods have been fully described in a previous paper[1] it will suffice to state here that the milk is first separated by means of a clay filter into two portions: the serum, which contains all the soluble constituents; and the unfilterable portion, which contains those substances which are present in the milk in the form of an emulsion or a colloid. These two portions are then studied separately.

RESULTS OF INVESTIGATIONS.

ACIDITY OF HUMAN MILK.

When the acidity is determined by the method recently published[2] it is found that the acidity of the serum of fresh milk is the same as that of the original whole milk, which proves that the serum contains all the acid bodies in the milk. It will be shown later that this acidity is in all probability due to acid phosphates and that from 3 to 6 c.c. of tenth normal alkali are required to make 100 c.c. of milk neutral to phenolphthalein.

In this connection it is well to call attention to the erroneous idea which has prevailed as to the difference in the acid reaction of human milk and cow's milk. This has been due to the incorrect method used to determine the acidity. The high acid figures previously obtained for cow's milk were due to the interference of the neutral calcium phosphate, $CaHPO_4$, which is present in

*Published in *The Journal of Biological Chemistry* 20: 707, 1915, as a contribution from the Boston Floating Hospital, Boston, and the Chemical Laboratory of the New York Agricultural Experiment Station, Geneva.

[1] L. L. Van Slyke and A. W. Bosworth: *Journ. Biol. Chem.*, 20:135, 1915; N. Y. Agr. Expt. Sta. Tech. Bul. No. 39, 1915.

[2] Van Slyke and Bosworth: *Journ. Biol. Chem.*, 19:73, 1914; N. Y. Agr. Exp. Sta. Tech. Bul. No. 37, 1914.

cow's milk but not in human milk, as will be shown later on in this paper. By using the method mentioned above it is found that 100 c.c. of cow's milk will require the addition of from 4.4 to 7.0 c.c. of tenth normal alkali to make it neutral to phenolphthalein. The acidity of human milk and cow's milk is thus seen to be practically the same, and the practice of adding lime water to modified cow's milk used for infant feeding as a means of correcting the acidity is thus shown to have no foundation.

THE SERUM OF HUMAN MILK.

The serum of fresh human milk resembles water in appearance. It is acid to phenolphthalein and alkaline to methyl orange, which would indicate that the acidity is due to acid phosphates. Its composition, in comparison with that of whole milk, is given in the following table.

TABLE I.—COMPOSITION OF HUMAN MILK AND ITS SERUM.

Constituents.	Original milk.	Milk serum 100 c.c.	Milk constituents in serum.
	Gm.	*Gm.*	*Per ct.*
Fat...............................	3.30	0.00	0.00
Casein.............................	} *1.20	{ 0.00	0.00
Albumin............................		0.131	*(13.10)
Nitrogen in other compounds calculated as protein.........................	0.307	0.307	100.00
Citric acid.........................	0.1055	0.1055	100.00
Phosphorus, organic..................	0.0008	0.00	0.00
Phosphorus, inorganic................	0.0148	0.0148	100.00
Calcium.............................	0.0354	0.0214	60.45
Magnesium...........................	0.0030	0.0030	100.00
Sodium..............................	0.0147	0.0147	100.00
Potassium...........................	0.0711	0.0711	100.00
Chlorine............................	0.0375	0.0373	100.00

*The determination of casein in the whole milk was very unsatisfactory, and for that reason the amount is not given in the table. It was about 0.2 gm. per 100 c.c. of milk.

THE UNFILTERABLE PORTION OF HUMAN MILK.

By reference to Table I it is seen that that portion of the milk which does not pass through the clay filter contains the fat, the casein, and part of the albumin and calcium. The fat can be removed by extraction with ether, leaving the protein and calcium. It was found impossible to separate the calcium from the protein by any mechanical means; so we conclude that it is in chemical combination with the protein. It is important to note that evidence seems to indicate that the albumin as well as the casein is combined with calcium. This will be investigated more carefully before any definite report is made.

The figures in Table I show that human milk, unlike cow's milk, contains no insoluble inorganic phosphates; *i.e.*, no dicalcium phosphate, $CaHPO_4$. This substance, if it were present, would be found in the portion of the milk which failed to pass through the clay filter.

PRINCIPAL COMPOUNDS OF HUMAN MILK.

As a result of our studies of human milk the following arrangement is offered as representing the probable condition in which the constituents are present in human milk of average composition.

	Per ct.
Fat	3.30
Milk sugar	6.50
Proteins combined with calcium	1.50
Calcium chloride	0.059
Mono-potassium phosphate (KH_2PO_4)	0.069
Sodium citrate ($Na_3C_6H_5O_7$)	0.055
Potassium citrate ($K_3C_6H_5O_7$)	0.103
Mono-magnesium phosphate ($MgH_4P_2O_8$)	0.027

My thanks are due to Dr. Henry I. Bowditch, who has used funds at his disposal to pay for the milk used for this investigation.

TECHNICAL BULLETIN No. 44.   AUGUST, 1915.

New York Agricultural Experiment Station.

GENEVA, N. Y.

ASCOCHYTA CLEMATIDINA, THE CAUSE OF STEM-ROT AND LEAF-SPOT OF CLEMATIS.

W. O. GLOYER.

PUBLISHED BY THE DEPARTMENT OF AGRICULTURE.

ASCOCHYTA CLEMATIDINA, THE CAUSE OF STEM-ROT AND LEAF-SPOT OF CLEMATIS.*

W. O. GLOYER.

SUMMARY.

(1) The stem-rot and leaf-spot of clematis are caused by the fungus *Ascochyta clematidina* (Thümen).

(2) The plants are killed by the growth of the fungus down the petiole into the stems, thus girdling the plant at the node. The stem may be girdled also by the lesions anywhere on the internodes. Dead stubs left on the vines are a means of holding the disease over a period of time. New shoots may be formed below the girdled region, but the downward progress of the fungus ultimately kills the plant if the diseased tissue is not removed.

(3) Overwintering out of doors does not kill the fungus in cultures or on dead vines. Whenever the temperature permits, the fungus resumes its growth.

(4) The fungus is readily isolated and grows well on the media generally employed in the laboratory.

(5) The disease has been successfully produced by inoculating *C. paniculata* and *C. jackmanni* with the mycelium from pure cultures. The fungus has been reisolated from such inoculations, and with it lesions were again produced on other vines similarly treated.

(6) *A. clematidina* is not related to other common species of the genus Ascochyta, for inoculations made in growing stems of bean, pea, muskmelon, pumpkin, eggplant, and the young shoots of elm gave negative results.

(7) Spraying the plants with spores will produce the leaf-spot. More spots are produced when the spores are placed on the lower surface of the leaf than on the upper. A temperature of 23.° C. is more favorable for the production of the leaf-spot than a temperature of 10° C.

(8) The matting of the vines produces a condition most favorable for the spread of the disease. Ventilation can be obtained by supporting the vines or by planting them far enough apart to prevent matting.

(9) On the hybrids the disease can be controlled in the forcing frames or in the greenhouse by the use of sprays. In the field, the spraying of hybrids properly supported is of little benefit.

* Also printed in *Journal of Agricultural Research*, 4: 331-342. 1915.

(10) On *C. paniculata* spraying with a fungicide checks the disease. In the field the removal of diseased leaves and vines before spraying is of practical value in controlling the disease.

(11) Sulphur dusted on *C. paniculata* in large quantities may cause injury.

(12) A mixture of 1 pound of laundry soap and 6 pounds of sulphur to 15 gallons of water, when sprayed on cuttings in the greenhouse or on *C. paniculata* growing in the beds, controlled the disease.

INTRODUCTION.

The sudden dying of clematis plants has been known for many years, and there has been much speculation as to its cause and prevention. Apparently the disease occurs in both Europe and America wherever the large-flowered kinds of clematis are grown extensively. From published accounts it is clear that the various writers had in mind the same disease, though they ascribed it to different causes. In 1884 Arthur (1) [1] studied a clematis stem-rot which he suspected of being due to the fungus *Phoma clematidis* Sacc. Trelease (16), Comstock (3), Klebahn (7), and others have considered nematodes as the causal agent. In specimens received from Klebahn, Ritzema Bos (2) found nonparasitic nematodes, while in material of his own collection he found the larvæ of a fly, *Phytomyza affinis*, and a species of Pleospora. Prillieux and Delacroix (9) and Morel (8) believed the disease to be of bacterial origin. Sorauer (13) reports a gall-like formation on the stem of *Clematis jackmanni* near the surface of the soil and attributes the death of the affected plants to *Gloeosporium clematidis*. Green (5, p. 284–285) has reported the relative susceptibility of a few varieties of clematis which he grew, but he did not attempt to ascertain the cause of the disease. Except for a preliminary abstract by the writer (4), the primary cause of the clematis disease has heretofore been unknown.

DESCRIPTION OF THE DISEASE.

The clematis disease manifests itself differently on the various species and hybrids. On hybrids grown in the field it is a stem-rot, while in the greenhouse, where the cuttings are propagated, it is a leaf-spot as well as a stem-rot. On *Clematis paniculata* the disease takes both forms.

C. paniculata, a type species, is propagated from seed [2] and when grown in uninfected cold frames or greenhouses remains free from

---

[1] Reference is made by number to " Literature cited," p. 14.

[2] For description of general methods of propagation and of the various species of clematis see the following:

Bailey, L. H., ed. Cyclopedia of American Horticulture. Ed. 7, 1: 327–332, fig. 485–492. 1910.

Le Bêle, Jules. The clematises. *Garden*, 53: 544–548, 54: 39–40, 127, 138, 200–201, 240–241. 1898. Translated from *Bul. Soc. Hort. Sarthe*, 1896.

disease. Such seedlings are either potted or placed in beds, where they are planted about an inch apart in rows 4 inches apart. In the fall, when these plants have made a growth of 8 to 10 inches, the leaf-spot may make its appearance and be thus carried over the winter on the dead leaves or in lesions formed on the vines. If these plants are left in the beds a second season, the fungus may make its appearance early in spring and increase until by midsummer no vine is wholly free from disease. The leaf-spot may first appear either as a mere dot or as a water-soaked area. With the advent of moist warm weather the former usually leads to the latter. On drying the water-soaked spot becomes tan-colored with a red margin. Plate I shows the general appearance of the disease on *C. paniculata*. The older leaves are badly diseased or dead, and the fungus has grown down the petiole to the node, where in time the vine may become girdled. The younger leaves show the early stages of the leaf-spot. The stem shows the lesions, reddish in color, formed at the nodes and on the internodes. Later these take on a gray color. Plate II illustrates a group of leaves of *C. paniculata* with spots that are zonate, owing to the unequal growth of the fungus under the influence of changes in temperature. The newly formed spot has a dark margin of red tissue and a lighter center. Pycnidia are produced on the diseased leaves. Succulent, growing tissues succumb more readily to the disease than do the woody stems. In the latter it may require a month for the fungus to pass a node. Plate III, fig. 1, shows a portion of a vine of *C. paniculata* 44 inches long on which the lower leaves were wilted, while the distal ones were still turgid. The fungus entered through the stub *a*. It girdled the stem and disintegrated the upper roots, leaving the central cylinder as the only means of communication with the healthy roots below. Pure cultures of *Asocochyta clematidina* were obtained from the boundaries of the lesion. Pycnidia were formed on the stem above the ground. In other cases pycnidia have been found on the epidermis, while the tissues underneath were healthy.

Some of the large-flowered kinds of clematis are grown from seed, but in America the majority of those cultivated are hybrids. They are propagated from cuttings taken from rapid-growing, disease-free vines. The cuttings are made in May or June and consist of a single node with the attached leaves and the internode below. They are placed in moist sand and exposed to bottom heat or else grown in forcing frames. In clematis forcing frames the humidity and temperature are usually higher than in the average greenhouse. Under these conditions, if the spores are present, a leaf-spot may be formed, and the entire cutting may be killed or the fungus may be halted at the node. The fungus that has been checked may again become active when the cuttings are potted and placed in the greenhouse, or new infections may take place on the leaves. In the fall some of the plants are placed in storage, while others are kept over winter in the greenhouse and the tops used for cuttings. In the following spring both lots are transplanted into the open field and, unlike *C. paniculata*, are not allowed to trail on the ground. Experience has taught the nurseryman that supported vines, owing to the better ventilation they receive, do not die as readily as those left on the ground. They make a vigorous growth, and yet when about to bloom they may suddenly die. It is at this stage that the disease first attracts the attention of the nurseryman, though in reality it was on the plants while they were still in the greenhouse and was there overlooked. Plate IV, fig. 1, shows a plant free from leaf-spot, yet girdled by the fungus lurking in the stub *a*, which in ordinary practice is not removed. Plate IV, fig. 2, is a reproduction of another vine of *C. jackmanni* that has many pycnidia of *A. clematidina* on the old stub. After the removal of this stub some of the discolored tissue still remained. The new shoot formed is wilting, and the split stem shows discolored fibrovascular bundles from which *A. clematidina* was isolated. In advanced stages the roots may disintegrate similarly to that shown in Plate III, fig. 1. The spots on the leaves of *C. jackmanni* resemble those found on *C. paniculata*, *C. recta*, and *C. virginiana*.

ISOLATION OF THE CAUSAL FUNGUS.

By previous writers the dying of clematis plants has been assigned to various factors, but none have discovered the primary cause. The dying of the leaves owing to lack of light, the breaking of the vine by strong winds, and injury by nematodes are factors that have been eliminated as primary agencies, while the constant association of *A. clematidina* points to it as the causal organism. The fungus can be readily isolated by the poured-plate method, using the spores from a crushed pycnidium, by the use of sterile leaf tissues, or by the use of free-hand sections of diseased material. The last-named method consists in making free-hand sections under as sterile conditions as possible by sterilizing the outer tissue and the instruments. If such sections show mycelium they are transferred to sterile media. Some have maintained that no mycelium can be seen in the decayed tissue, but the writer has observed in the tissues 3 to 5 mm. from the boundaries of the lesions mycelium which in plate cultures proved to be that of the causal organism, *A. clematidina*.

A. clematidina grows well on the media generally used in the laboratory. It grows at about the same rate on nutrient glucose agar, oatmeal agar, bean pods or stems, moist oats, and corn meal; producing pycnidia in five to seven days. These pycnidia may show a pink tinge at first and later turn brown. The fungus grows less vigorously on corn-meal agar, potato agar, starch agar, sugar-beet plugs, apple twigs, and sterile raw carrot. Oatmeal and starch agar are at first turned green, but later take on a brown color. On starch agar the mycelium penetrates the medium and forms chlamy-dospores, as shown in Plate III, fig. 1. These are thick-walled, green-brown bodies filled with oil globules. When placed in water, they germinate readily.

INOCULATION EXPERIMENTS.

To prove the pathogenicity of *A. clematidina*, mycelium from pure cultures was inoculated into stems of *C. paniculata* and *C. jackmanni*. In all cases lesions were produced, while the checks remained normal. From such lesions the fungus was reisolated, and, when again inoculated into either host, typical lesions were produced. In all, four sets of inoculation experiments were carried out at various times, making from 3 to 10 inoculations on each of 32 plants. Inoculations on succulent stems caused the vines to wilt in four days, while in one case an inoculation on a woody vine 6 mm. in diameter required 47 days to kill the plant. Pycnidia were produced on all lesions.

Plants of *C. paniculata* were sprayed with sterile water containing spores of *A. clematidina* and then kept under bell jars for two days. on the third day the leaves showed water-soaked spots of various sizes, while the checks, which had been sprayed with sterile water, remained free from disease. To test the effect of temperature on infection, two plants were sprayed with the same spore-laden water and then subjected to different temperatures — 23° C. and 10° C. At the end of five days the plant kept at 23° showed 45 leaf spots, while the plant kept at 10° showed but 1 spot.

Spores placed on the lower surface of the leaves produced more spots than those placed on the upper surface. Typical lesions were also produced. on the roots by inoculating them with the mycelium from a pure culture.

The *A. clematidina* isolated from *C. paniculata* was inoculated into growing stems of bean, pea, muskmelon, pumpkin; into stems, petioles, and fruits of eggplant (var. Black Beauty); and into the young shoots of elm. In all cases negative results were obtained. On most of these plants pycnidia were produced on the tissues killed in making the wound, but in no case did the mycelium penetrate the healthy tissues and form a lesion.

TAXONOMY OF THE FUNGUS.

Arthur (1) observed a species of Phoma, possibly *P. clematidis*, on clematis, but on consulting the original notes made by him it is clear that he had a fungus different from that found by the writer. On but one occasion has *Phoma* sp. been found and that was a saprophyte on the leaf of *C. paniculata*. It was isolated in pure culture, the mycelium inoculated into the stems, and the spores sprayed on leaves, but in no case were lesions or leaf-spots produced.

Saccardo (11) notes *A. clematidina, A. vitalbae, A. indusiata*, and
A. davidiana as occurring on various species of clematis, and their
chief point of difference is in the size of the spores. The writer has
examined the specimens of *A. clematidina* Thümen on *C. virginiana*
collected by Mr. J. J. Davis in Wisconsin and distributed in Fungi
Columbiani No. 2503; also those of *A. indusiata* Bres. on *C. recta*
in Krieger's Fungi Saxonici No. 1189. In both, the spots resemble
those found on *C. paniculata* and *C. jackmanni*. In the former the
spores are cylindrical 1-septate and hyaline. They measure 8 to
12 by 3.2μ, the average dimensions being 9.5 by 3.2μ. The spores
of the latter species are hyaline to honey-colored, somewhat constricted, and measure 12 to 22 by 6.3μ, with an average of 19
by 6μ.

The writer has repeatedly examined the species of Ascochyta on
clematis and found it quite variable. The chief difference is in the
spores, though sometimes the pycnidia are more deeply immersed
than at other times. Plate V, fig. 1, shows a pycnidium in the leaf
tissues of *C. paniculata*. The spores vary in length from 6 to 28μ
and in width from 3 to 6μ, but generally they are about 9 to 13 by
3 to 4μ. Plate III, fig. 3, shows the typical spores. The spores are
either 1- or 2-celled, rarely 3-celled. Some leaves of *C. jackmanni*
collected in the fall of 1914 showed pycnidia having spores as long
as 28μ and averaging 18 by 5.7μ. Inoculations with material from
cultures obtained by the isolation of single spores showed that this
fungus was the same as that usually encountered. The various
differences in color shown by the spores disappear when the spores
are plated out under control conditions. Considering the varia-
bility of the fungus found by the writer, any of the descriptions
given for the different species of Ascochyta described on clematis
would in general apply to it. Hence, the name selected is the
oldest one, *Ascochyta clematidina* Thümen, the description of which
is here emended as follows:

Ascochyta clematidina (Thümen).

Ascochyta clematidina Thümen, Pilzfl. Sibir. n. 619, 1884, *in* Sacc. Syll. Fung.,
v. 3, p. 396.

On stems and foliage; spots circular, zonate to indefinite; pycnidia (on leaves mostly
epigenous, sometimes hypogenous) tan to dark brown, scattered to gregarious, globose
to subovoid, immersed, then erumpent, ostiolate, averaging 120μ in diameter; spores
variable, subellipsoidal to cylindrical, 1- or 2-celled, septa more or less medial, sometimes constricted, hyaline to dilute honey or olive color, often guttulate, 6 to 28 by
3 to 6.4μ, usually 9 to 13 by 3 to 4μ; exuded spore mass honey-colored, sometimes
pink.

On living leaves and stems of *Clematis paniculata, C. virginiana*, and the hybrids
*C. hendersoni, C. henryi, C. jackmanni, C. ramona, C. Duchess of Edinburg, C. Mme.
Baron Veillard*, and *C. Mad. Édouard André*. According to Von Thümen, it occurs
also on living leaves of *C. glauca*. As yet no perfect form of *A. clematidina* has positively been found.

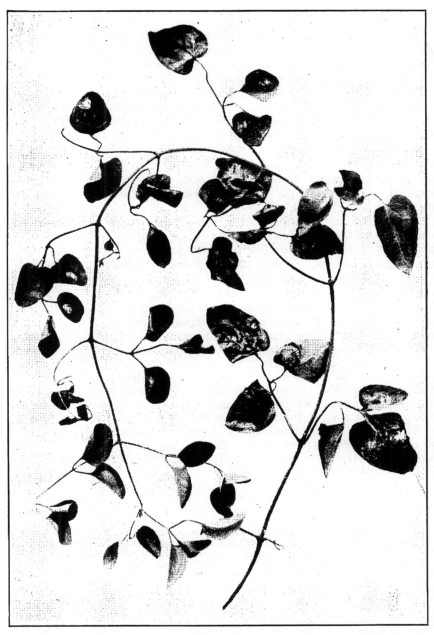

PLATE I.— *Clematis paniculata:* PORTION OF VINE SHOWING GENERAL NATURE OF LEAF SPOT.

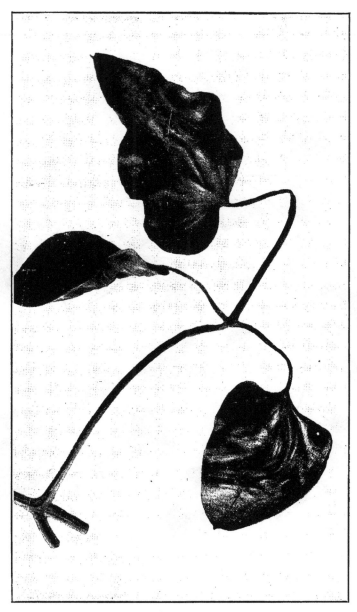

PLATE II.—*Clematis paniculata:* GROUP OF LEAVES ENLARGED TO SHOW ZONATION AND PYCNIDIA OF SPOTS. ONE LEAF SHOWS NEWLY FORMED SPOT, WITH LIGHTER CENTER.

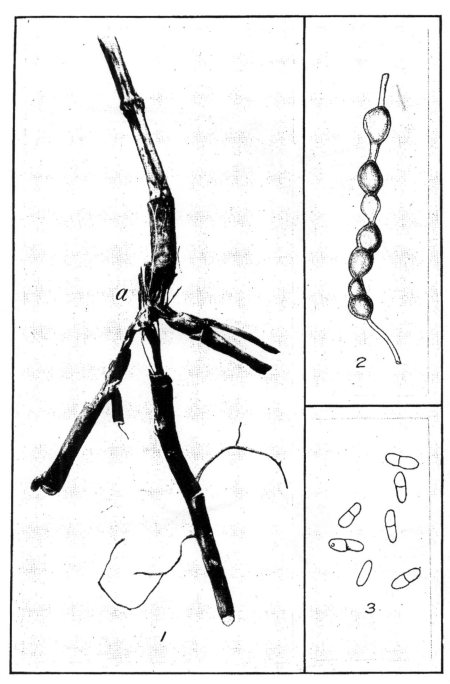

PLATE III.
(For explanation, see back of Plate V.)

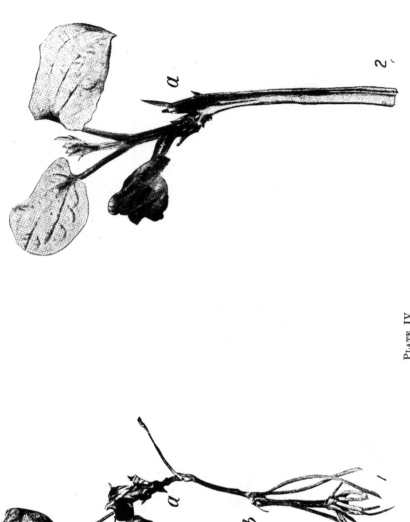

PLATE IV.

(For explanation, see back of Plate V.)

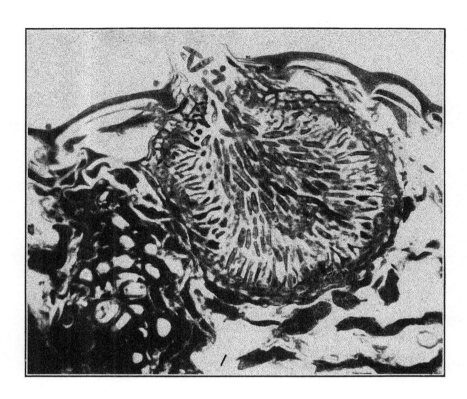

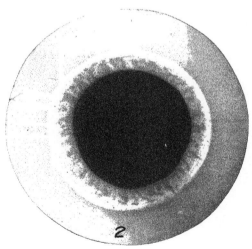

PLATE V.
(For explanation, see back of this plate.)

EXPLANATION OF PLATES III, IV AND V.

PLATE III.

Fig. 1.—*Clematis paniculata:* A portion of a vine 44 inches long that showed indications of wilting of the lower leaves while the distal leaves were still turgid. The fungus entered through the stub *a*. In the girdled region the parenchyma of the roots had disintegrated, leaving the central cylinder as the only means of communication with the healthy roots below. *Ascochyta clematidina* was isolated in pure cultures from the boundaries of the lesion.

Fig. 2.—*Ascochyta clematidina:* Chlamydospores as formed on starch agar.

Fig. 3.—*Ascochyta clematidina:* Camera-lucida drawing of spores.

PLATE IV.

Fig. 1.—*Clematis jackmanni:* A vine free from leaf-spot that has been girdled by *Ascochyta clematidina* in the region of the previous year's stub *a*. A new shoot would have been sent forth from an active bud at *b*, but it would have soon died, for the fungus had discolored the vascular bundles beyond this point. The presence of the fungus was proved by isolating it from the discolored tissue.

Fig. 2.— *Clematis jackmanni:* Plant from which the diseased stub *a* was cut away without removing the discolored tissue. The leaves were free from leaf-spot and were drying. The split stem shows the discolored fibrovascular bundles from which the fungus was isolated.

PLATE V.

Fig. 1.—*Ascochyta clematidina:* Photomicrograph of a pycnidium from stained section of a leaf of *Clematis paniculata*.

Fig. 2.—*Ascochyta clematidina:* Culture growing on agar to which Ivory soap at the rate of 1 pound to 15 gallons of water was added, showing the oily film about the margin of the culture in which the crystals of stearic acid are found.

CONTROL EXPERIMENTS IN 1913.

In 1913 some 2-year-old plants of *C. paniculata* that had made a dense, matted growth of tangled vines were badly diseased, while a bed of seedlings next to them was free from disease. In an attempt to save the 2-year-old plants, they were cut back to a length of 4 to 6 inches and then sprayed with bordeaux mixture on July 21. A small area was left unpruned and unsprayed as a check. By October 17 the seedlings, which had made a growth of 8 to 10 inches, showed an occasional leaf spot. The pruned-and-sprayed plat produced an excellent growth, but had some leaf-spot and a few girdled vines. The check showed many dead plants, and none of the living ones were entirely free from disease.

CONTROL EXPERIMENTS IN 1914.

A writer in *Garden* (10) states that bordeaux mixture, when applied to diseased clematis plants, was of no benefit in checking the disease. In 1914, spraying experiments were carried out by the writer on 18 rows (about 300 feet long) of plants of *C. jackmanni*, half of which were sprayed with bordeaux mixture (4–4–50 formula), while the others were left as checks. Four of these, two checks and two sprayed rows, were pruned on June 12 and 25 in such a manner as to remove the dead stubs of the previous year. Plants from which all of the discolored tissue could not be removed without injury to the entire vine were marked with tags. The rows receiving bordeaux mixture were sprayed every two weeks. The final examination was made on October 19. No difference could be seen between the sprayed and the check rows either in the amount of leaf spot or the number of dead plants. The same held true for the pruned and unpruned rows. However, there was but little leaf-spot, and it was observed that the dead plants in the pruned rows were invariably plants that had been tagged. No doubt the pruning was done too late in the season to be of any benefit. Sulphur dusted on cuttings in the forcing frames did not check the disease. Plants in the greenhouse sprayed with soap-and-sulphur mixture so as to cover the leaves with a thin film were healthier than the unsprayed plants. These, however, were not carried through the second season, and hence the ultimate results are unknown.

Two long, narrow beds of *C. paniculata* were utilized for spraying and dusting experiments in 1914. Bed 1 consisted of yearling plants untreated in 1913. Bed 2 contained 2-year-old plants pruned and sprayed with bordeaux mixture in 1913. Both beds were divided into plats 6 by 25 feet in size.

Two plats in each bed were sprayed with bordeaux mixture six times at intervals of two weeks from May 15 to August 8. On the same dates one plat in each bed was sprayed with a soap-and-sulphur mixture composed of 1 pound of soap, 6 pounds of sulphur, and

15 gallons of water. Two and one-half gallons of the mixture, containing 1 pound of sulphur, were used on each plat at each application. On two plats in bed 1 and one plat in bed 2 the plants were dusted six times with sulphur, using 1 pound to the plat at each application. The remaining eight plats (three in bed 1 and five in bed 2) were left untreated for checks.

As the season advanced, the virulence of the disease increased, becoming quite severe on all three check plats in bed 1 and one check plat in bed 2. On August 8 some of the plats were pruned, thus terminating the main experiment. The effect of the different kinds of treatment up to August 8 is shown in Table I.

TABLE I.—RESULTS OF AN EXPERIMENT ON THE CONTROL OF LEAF-SPOT AND STEM-ROT OF *Clematis paniculata*.

Bed No. 1.			Bed No. 2.		
Plat No.	Treatment.	Percentage of plants free from disease.	Plat No.	Treatment.	Percentage of plants free from disease.
1	Bordeaux mixture.....	75	9	Bordeaux mixture.....	60
2	Check...............	2	10	Check...............	2
3	Bordeaux mixture.....	80	11	Sulphur.............	70
4	Check...............	10	12	Check...............	45
5	Sulphur.............	80	13	Soap and sulphur......	100
6	Soap and sulphur......	100	14	Check...............	85
7	Check...............	10	15	do	85
8	Sulphur.............	65	16	do	85
			17	Bordeaux mixture.....	95

The results are more uniform in bed 1 than in bed 2. This may be due to the treatment of bed 2 the previous year. That there was more disease in plats 9 and 10 than in the other plats of bed 2 may be accounted for by the fact that these two plats were used as checks in 1913, and hence were neither pruned nor sprayed in that year. Plats sprayed with the soap-and-sulphur mixture remained free from leaf-spot and lesions on the stems; hence, their condition is rated at 100 per ct. In rating the other plats the amount of leaf-spot, the number of lesions, and number of girdled plants have all been considered.

On the plats 1, 3, and 17, which were sprayed with bordeaux mixture and which were not pruned on August 8, the spraying was continued to the end of the season. Check plats 4, 14, and 16 also were left unpruned. On October 19, when the final examination of these plats was made, it was found that in plats 1 and 3 the leaves on the new growth were disease-free and that there was but an occasional dead vine. On check plat 4 half of the newly formed leaves were diseased, and about one-third of the vines were dead. Plats 16 and 17 showed about the same amount of leaf-spot, only an occasional spot. The former, however, showed more lesions on the stems than the latter. Check plats 2, 7, 10, and 12, which had been pruned back to a length of 4 to 6 inches and then given one application of bordeaux mixture, had but little disease as compared with plat 4, which had received no treatment whatsoever.

INJURIOUS EFFECTS OF SULPHUR ON CLEMATIS PANICULATA.

The promising results obtained by Smith (12) in dusting asparagus with sulphur for the control of rust led the writer to try sulphur in controlling the fungus on clematis. Up to August 8 the results were satisfactory, and no injury was observed. Soon after the pruning of August 8 there were several hot, dry days followed by a period of rainy weather, during which water accumulated at the end of bed 1. Up to the end of the season only one plant in plat 8 had sent forth a new shoot. The other vines were dead, and the stems at the surface of the ground for about an inch were discolored. A particle of soil placed on the tongue had an acid taste. According to a test made by Mr. R. F. Keeler, 1 gm. of this soil is equivalent to 0.5 c.c. of $\frac{N}{10}$ acid, while soil from the adjoining check plat 7 was neutral. In check plat 7 a few vines died, owing to the lack of drainage, but it seems apparent that in the other cases the injury was caused by sulphur that had washed from the foliage and had accumulated in the upper layer of soil. As the season advanced, sulphur injury was observed in the other treated plats, but in these cases the injury was localized in areas not larger than 2 feet in diameter. The injury began to show on the plats sprayed with the soap-and-sulphur mixture after nine applications had been made, while in the plats dusted with sulphur it appeared after six applications had been made.

SOAP AND SULPHUR AS A SPRAY MIXTURE.

A mixture of about 1 pound of laundry soap and 6 pounds of sulphur in 15 gallons of water was in common use as a greenhouse spray at the nursery where the spraying experiments were conducted. It was used with success in the control of leaf-blotch, *Diplocarpon rosae* Wolf, on susceptible varieties of roses grown in the forcing houses. Halsted and Kelsey (6) used Ivory soap at the rate of 1 ounce to

4 gallons of water for spraying *Phlox drummondii* and the common verbena attacked by powdery mildew and were able to check it. Another (15) has shown that soap at the rate of 1 ounce to 1 gallon of water controlled the mildew and aphids of roses. R. E. Smith (12) recommended that, in the absence of dew, whale-oil soap be sprayed on asparagus tops to hold the sulphur that is to be dusted over them for the control of the rust. Spieckermann (14) has shown that weak solutions of soap have a nutritive value and can be assimilated by the higher fungi.

In order to test the toxic effect of soap, mycelium of *A. clematidina* was transferred to petri dishes containing soap agar of different strengths — viz, 2 per ct. agar containing alkali-free Ivory soap in the proportion of 1 pound to 5, 10, 15, 20, and 40 gallons of the medium. Fifteen c.c. of such media were placed in each petri dish. When the fungus became established, the diameter of the culture was measured daily, and the rate of growth was considered as a measurement for toxicity. Cultures on 2 per ct. agar and nutrient-glucose agar served as checks. Table II gives the averages of growth of four or five cultures on each medium grown under the same conditions at room temperature.

TABLE II.—SUMMARY OF THE DATA ON THE TOXIC EFFECT OF SOAP AGAR OF VARIOUS STRENGTHS ON *Ascochyta clematidina*.

Culture medium.	Number of cultures.	Average diameter of cultures after growth for—									
		3 days.	5 days.	7 days.	9 days.	11 days.	13 days.	15 days.	17 days.	20 days.	23 days.
		Mm.	*Mm.*	*Mm.*	*Mm.*	*Mm.*	*Mm.*	*Mm.*	*Mm.*	*Mm.*	*Mm.*
Soap agar (1 lb. to 5 gals.)..	5	0	0	0	0	0	0	0	0	0	0
Soap agar (1 lb. to 10 gals.).	5	0	0	0	0	0	0	0	0	0	0
Soap agar (1 lb. to 15 gals.).	4	0	0	7	9	11	15	18	20	22	28
Soap agar (1 lb. to 20 gals.).	5	0	11	15	19	23	26	30	34	39	44
Soap agar (1 lb. to 40 gals.).	4	.8	20	25	32	40	44	50	56	67	73
Check, 2 per cent agar....	5	10	21	29	40	51	61	70	77	85	....
Check, nutrient-glucose agar..............	5	15	29	38	48	60	69	77	...	...	....

Plate V, fig. 2, shows a culture of *A. clematidina* on soap agar. The concentric ring or oily film about the culture may be 2 to 7 mm. wide. Upon the drying of the medium, crystals of stearic acid are seen only in this region, indicating that an active principle given off by the fungus liberated the fatty acid. The culture shows a green discoloration, which is due to the formation of brown-green, thick-walled chlamydospores. These bodies, as well as the mycelium, are filled with oil globules, while none are found in the 2 per ct. agar. These cultural experiments indicate that soap at the strengths in which it is used as a contact insecticide has in itself fungicidal value, as well as being a means of adhesion or suspension for other materials.

METHODS OF CONTROLLING THE FUNGUS

The suggestions here given for controlling *A. clematidina* are based upon the observations and experiments made in the last three years. Greater success can be attained by changing the methods of culture than by spraying. Long experience has taught the nurseryman that there is less disease when the hybrids are supported while growing in the field or in the greenhouse than when they are permitted to trail on the ground. This holds true also for *C. paniculata*, but its selling price does not warrant so much expense for labor. This can be overcome by transplanting the plants from the beds to the open field after the first year, placing them far enough apart to prevent matting. Spraying is beneficial to such plants, but before making such applications it is advisable to remove all diseased leaves and dead vines. Plants so treated are disease-free in the fall. If seedlings are grown in a greenhouse where clematis has never been grown before and are kept away from older diseased beds, they will remain disease-free. The fungus can live as a saprophyte on dead vines kept out of doors in baskets, and under such conditions it has lived over two winters, producing pycnidia and viable spores in abundance. This indicates that the same beds should not be used for clematis in successive years.

On hybrids the disease is primarily a greenhouse trouble and can be overcome by the use of cuttings made from healthy plants. A light spraying with the soap-and-sulphur mixture has proved satisfactory in the greenhouse. It could readily be applied also in the forcing frames. Diseased leaves or stubs should be removed as soon as discovered so as to prevent the fungus becoming established in the tissues.

The retail purchaser of clematis can prevent the dying of plants by taking proper simple precautions. The plants should be placed in good soil, well drained and on a sunny exposure. As soon as the new shoots have formed the old vine tissue should be carefully cut away close to the new shoots, removing all traces of the brown, discolored wood in which the fungus is to be found. Proper ventilation is obtained by training the plants to a strong trellis.

14

LITERATURE CITED.

(1) ARTHUR, J. C.
    1885 Disease of clematis. N. Y. State Agr. Exp. Sta. 3d Ann. Rpt., 1884, p. 383–385, fig. 5.

(2) Bos, J. Ritzema.
    1893. Phytomyza affinis Fall., as a cause of decay in clematis. *Insect Life* 6:92–93.

(3) COMSTOCK, J. H.
    1890. The clematis disease, or nematodes infesting plants. West N. Y. Hort. Soc. Proc. 35th Ann. Meeting, 1890, p. 7–12.

(4) GLOYER, W. O.
    1914. Stem rot and leaf spot of clematis. *Phytopath.* 4:39.

(5) GREEN, S. B.
    1906. Ornamental trees, shrubs and herbaceous plants in Minnesota. Minn. Agr. Exp. Sta. Bul. 96, p. 241–350, 108 fig.

(6) HALSTED, B. D., and KELSEY, J. A.
    1903. Fungicides and spraying. *In* N. J. Agr. Exp. Sta. 23d Ann. Rpt., [1901] /02, p. 415–417, pl. 10–11.

(7) KLEBAHN, Heinrich.
    1891. Zwei vermutlich durch Nematoden erzeugte Pflanzenkrankheiten. *Ztschr. Pflanzenkrank.*, 1:321–325.

(8) MOREL, F.
    1903. La maladie noire des clématites. *Rev. Hort.*, 75:364–365.

(9) PRILLIEUX, E. E., and DELACRIOX, GEORGES.
    1894. Maladies bacillaries de divers végétaux. *Compt. Rend. Acad. Sci.* [Paris], 118:668–671.

(10) S., E.
    1898. Clematises dying. *Garden*, 54:99.

(11) SACCORDO, P. A.
    1884–1906. Sylloge Fungorum . . . v. 3, p. 396, 1884; v. 10, p. 301, 1892; v. 14, p. 942, 1899; v. 18, p. 347–348, 1906. Patavii.

(12) SMITH, R. E.
    1906. Further experience in asparagus rust control. Cal. Agr. Exp. Sta. Bul. 172, 21 p., 7 fig.

(13) SORAUER, PAUL.
    1897. Zur Kenntnis der Clematis-Krankheiten. *Ztschr. Pflanzenkrank.*, 7:255–256.

(14) SPIECKERMANN, ALBERT.
    1912. Die Zersetzung der Fette durch höhere Pilze. I. Der Abbau des Glycerins und die Aufnahme der Fette in d e Pilzzelle. *Ztschr. Untersuch. Nahr. u. Genussmtl.*, 23:305–331, 3 pl.

(15) T., J. C.
    1898. Soft soap for mildew and insects. *Garden*, 53:492–493.

(16) TRELEASE, William.
    1885. Root-galls caused by worms. *Cult. & Country Gent.*, 50:354.

TECHNICAL BULLETIN No. 45. AUGUST, 1915.

New York Agricultural Experiment Station

GENEVA, N. Y.

INHERITANCE OF CERTAIN CHARACTERS OF GRAPES.

U. P. HEDRICK AND R. D. ANTHONY.

PUBLISHED BY THE DEPARTMENT OF AGRICULTURE.

INHERITANCE OF CERTAIN CHARACTERS OF GRAPES.*

U. P. HEDRICK AND R. D. ANTHONY.

SUMMARY.

In the last 25 years, during which time nearly 10,000 seedlings have been grown, various changes in the methods used at the Geneva Experiment Station have been made as the knowledge of breeding laws has been extended.

Results have compelled the belief that improved varieties of grapes will not be produced to any extent until the fundamental laws of heredity are understood. The present aim is to discover these laws.

The work is now progressing mainly along two lines: (1) The determination of the breeding possibilities of varieties of grapes and (2) the study and interpretation of breeding phenomena.

Nearly 200 varieties of grapes have been used in the breeding work.

Much of the value of the early work was lost by growing too few seedlings of each cross.

Recently *Vitis vinifera* has been used to a considerable extent in hybridization.

The usual method of emasculation has been ineffective in a few cases and may be open to criticism.

One of the surprises in the study of grape varieties was the failure of many commercial sorts to transmit desirable qualities.

In order to study grape varieties, nearly 3,000 selfed, or pure, seedlings have been grown. These are uniformly lacking in vigor.

The inheritance of those grape characters which have sufficient data available has been discussed in this paper.

Of the two types of stamens, reflexed and upright, the first is correlated with complete, or nearly complete, self-sterility, the second with self-fertility.

Self-sterility is probably caused by impotent pollen. It exists in varying degrees and depends to some extent upon the condition of the vine and environmental factors.

---

* Also printed in *Jour. Agr. Research*, 4:315–330, 1915.

Self-sterile varieties of grapes being undesirable from a horticultural standpoint, can we eliminate those with reflexed stamens? The following crosses have given upright and reflexed stamens in the indicated ratios:

$$U \times U = 4.3 \ U:1R$$
$$R \times R = 1.2 \ U:1R$$
$$R \times U = 1U:1R$$
$$U \times R = ?$$

Breeding from the upright varieties only will decrease but not eliminate the seedlings with reflexed stamens.

The following seem to be the results secured in the study of sex inheritance:

Hermaphrodite female × hermaphrodite male = all hermaphrodites.

Hermaphrodite female × pure male = $\frac{1}{2}$ hermaphrodites + $\frac{1}{2}$ males.

Two general laws have been formulated with regard to color of the skin: (1) White is a pure color; and (2) it is recessive to both black and red.

No black variety has proved pure for blackness. Some contain white, others white and red. Red varieties are equally diverse.

The colors of pure seedlings of certain varieties show wide variation, even when derived from varieties of similar color.

In the inheritance of quality the most noticeable thing is the low percentage of seedlings whose quality is good or above good. This is probably due to the leveling influence of the wild ancestors from which the seedlings are but a step removed.

Most grapes of high quality possess some *V. vinifera* blood. This predominance of high quality is probably due to the intense selection to which the species has been subjected for centuries.

The pure seedlings on the New York Station grounds have been lower in quality than crossed seedlings.

In the inheritance of size of berry there is no indication of dominance of any one size, though there is a tendency for a variety to produce seedlings approaching its own size.

The oval form of many *V. vinifera* hybrids is probably an intermediate between round and a more pronounced oval. Oblate may be a pure form recessive to round.

The season of ripening of the parent influences to a considerable extent the season of the offspring.

A vineyard of 1,500 seedlings bred from 1898 to 1903 has dwindled through selection to less than 75. Of these, 5 have already proved promising enough to be named.

INTRODUCTION.

The breeding of grapes (*Vitis* spp.) was begun in the Horticultural Department of the New York Agricultural Experiment Station about 25 years ago and has been continued throughout this time as a horticultural problem. Nearly 10,000 seedlings have been grown, and of these about 6,000 have fruited. This work was begun about 1885 by Prof. E. S. Goff, who at first grew seedlings and plants from seeds open to cross-pollination. Later he crossed a number of varieties. In 1891 Prof. S. A. Beach became the Station horticulturist, and besides seeking to obtain new varieties he made studies of self-sterile varieties, studied the correlation between the size of seeds and vigor of plants, and did considerable hybridizing. In 1905 the senior author took charge of the work in horticulture at the Station. Mendel's work had just been discovered, and plant breeding was undergoing a stimulus from it. The work with grapes was therefore replanned and extensively added to with a view to studying problems of inheritance. This work has been continued and increased from year to year. Several assistants and associates have spent much of their time working with grapes at this Station. Mr. N. O. Booth worked with grapes from 1901 to 1908; Mr. M. J. Dorsey from 1907 to 1910; Mr. Richard Wellington from 1906 to 1913; Mr. R. D. Anthony, the junior author, began work at this Station in the summer of 1913, and has devoted most of his time to the grape work since then. Upon him has fallen the task of presenting the data in this paper. It is the purpose of this paper to discuss certain results of this work.

AIMS, METHODS, AND MATERIALS.

During this quarter of a century, experience and a better understanding of the principles of breeding have modified many of the methods and changed considerably the nature of the data which are now taken.

The ultimate aim in this work is, of course, the production of improved horticultural varieties. Through the early days, when breeding laws and methods were less understood than now, there was a tendency to make this the immediate as well as the ultimate aim. The fact that the first twenty years of grape breeding produced but one variety worthy a name served to confirm the conviction that this goal would be reached quicker by forgetting it for the time being and bending every effort to the discovery of how grape characters are transmitted.

The work is now proceeding mainly along two lines: (1) The determination of the breeding potentialities of a considerable number of varieties of grapes, especially with the view of finding unit characters; and (2) a review of all the New York Experiment Station breeding data on this fruit, to study and interpret breeding phenomena, accompanying this review with the making of the crosses necessary to throw further light upon doubtful points.

The results secured in testing the breeding possibilities of grape varieties, which will be discussed later, have made it seem desirable to extend this study to all the varieties which show any promise. For this reason nearly 200 different kinds have been used as pollen parents and nearly 100 as maternal parents in this work.

Frequently during the early days of the work seedlings which seemed to lack vigor in the nursery were discarded, instead of being planted in the test vineyards. Though this undoubtedly removed many unpromising seedlings, it seriously decreased the number which fruited, and made the interpretation of results difficult and uncertain. At best the number of seedlings that lived of each cross was smaller than could be desired, and, when this number was still further decreased by selection in the nursery or by untoward circumstances, much of the value of the work from a breeding standpoint was lost.

Another change of method which bids fair to be exceedingly important has been the use of varieties of *Vitis vinifera* in breeding. Every indication points to the desirability of the addition of some of the blood of the European grape to our native sorts. Although we are working primarily to determine breeding laws, there is usually a wide choice of varieties which answer our purpose, and with the growing of nearly 100 varieties of this species on the New York Station grounds we have been able to use several as parents. There are now several hundred hybrids containing *V. vinifera* blood growing on the Station grounds, and these will be increased by many hundreds during the following years.

The methods used in the actual work of crossing are similar to those of most breeders. The female blossoms are emasculated before the calyx cap splits off and are then bagged; the male blossoms are also bagged before the calyx splits. When the pollen is ripe, the bagged male cluster is usually cut from the vine and all or part of it brushed over the emasculated female. Usually some of the male cluster is inclosed in the bag, which is again put over the female after pollination. In a few cases, where the periods of blossoming of two varieties are too widely separated, it has been necessary to save pollen in clean glass jars. It is customary to dip the forceps used in pollination into alcohol with each new variety.

Certain results secured in the summer of 1914 seem to indicate that this method is perhaps open to criticism. While emasculating clusters of the Janesville variety it was found that, although the cap had not split, the pollen in the anthers seemed to be mature, and, as the anthers were ruptured during the emasculation, there was a possibility of self-fertilization taking place. Several clusters were emasculated and bagged without being pollinated. These set nearly the full quota of berries with seeds that have every appearance of being viable. With two other varieties, clusters emasculated and not pollinated matured a few plump seeds, though the clusters were much below normal. A somewhat similar instance is reported by Beach (1),[1] the variety being the Mills. This point deserves careful study, for if it is found that serious danger of self-pollination exists before the calyx cap splits it will be necessary to change the method — at least to the extent of emasculating the clusters several days before the cap is ready to come off.

All data regarding size and shape have been recorded in comparative terms, instead of the actual measurements being taken. With a limited number of observers and thousands of seedlings of the various fruits to be studied each year, it was a physical impossibility to take measurements and would not have increased the value of the records to any extent from a horticultural point of view, though it would, of course, have furnished interesting material for a statistical study. The value of data reported in comparative terms depends upon the accuracy of the recorder. The work with the grape has always been done by members of the scientific staff, and the observations have usually been checked during several seasons.

GENERAL RESULTS OF THE STUDY OF VARIETIES OF GRAPES.

One of the surprises in the study of varieties of grapes was the failure of many of our commercial sorts to transmit desirable qualities to their progeny. Seedlings of Concord, Niagara, Worden, Delaware, and Catawba grapes have so far proved only disappointments. The best results have been secured from such little-grown varieties as the Ross, Collier, Mills, Jefferson, Diamond, and Winchell. This has made it seem desirable to test all varieties that show any promise. The first step, then, was to secure as many varieties as possible which were of any value and which could be grown in northeastern United States. More than 400 such varieties have fruited in the Station's vineyards and have been described. About 200 of these have been used to a greater or less extent in the breeding work.

As an aid in studying the breeding possibilities of grape varieties the Station has grown nearly 3,000 selfed, or pure, seedlings, using as parents most of those varieties which have entered into the crosses that have been made. These seedlings have thrown much light upon the inheritance of various factors, but they have been so uniformly lacking in vigor as to lead to the belief that only through crossbreeding can we hope to produce improved varieties.

---

[1] Reference is made by number to " Literature cited," p. 19.

INHERITANCE OF CHARACTERS.

The grape characters discussed in the following pages are those for which sufficient data are available to make such a discussion of value.

SELF-STERILITY.

On the basis of flower type, grapes may be divided into three classes: (1) True hermaphrodites; (2) hermaphrodites functioning as females, owing to completely or partially abortive pollen; and (3) pure males with the pistil absent or rudimentary. Among these classes there are two types of stamens: Those with upright filaments and those in which the filaments bend backward and downward soon after the calyx cap falls off. According to Dorsey (5), this is due to the cells of the outer surface of the reflexed stamens being smaller and having thinner walls.

So far as observed at the New York Station, all pure males have upright stamens. Among the two classes of females Beach (1) found that only those varieties with upright stamens were capable of producing marketable clusters of fruit when self-fertilized. At the same time he reported nine varieties with upright stamens to be self-sterile. Since Beach published his report further work at this Station with three of these nine varieties has proved them self-fertile, and it is probably safe to say that all varieties with upright stamens are self-fertile, though in varying degrees. Reflexed stamens are always correlated with complete or nearly complete self-sterility. Reimer and Detjen (6) found this last conclusion to hold also with *Vitis rotundifolia*, a species not studied at this Station.

The cause of self-sterility in varieties with reflexed stamens seems to be a lack of viability in all or a larger part of the pollen of such varieties. Booth (2) found that such pollen was quite irregular in form and would not germinate in sugar solutions. Reimer and Detjen (6) state " the pollen of all the present cultivated [female] varieties [of *V. rotundifolia*] is worthless." Recently Dorsey (4) has ascribed the cause of this self-sterility to a degeneration in the generative nucleus. While this impotency may be absolute in many of the varieties, in some at least it is only relative. Frequently viable pollen will be found mixed with the usual misshapen, abortive pollen of the self-sterile varieties, and nearly a hundred pure seedlings of the varieties which Beach (1) reported as totally self-sterile have been grown in the Geneva Station vineyard. The degree of sterility seems to depend to some extent upon the condition of the vine due to environmental factors.

From a practical standpoint it is undesirable to grow self-sterile varieties. They will not succeed in the large blocks of the commercial plantation, nor are they always properly fertilized in the small home vineyard. Can we, then, in our grape breeding eliminate self-sterility? Letting U stand for upright stamens and R for reflex, the following table gives the results of our crosses:

U × U* = 180U + 47R	R × R = 16U + 16R	R × U = 207U + 206R
U selfed = 691U + 152R	R selfed = 94U + 73R	Ratio.........1U:1R
		U × R =?
Total....871U + 199R	Total....110U + 89R	
Ratio.........4.3U:1R	Ratio.....1.2U:1R	

Of the varieties whose pure seedlings have entered into the ratio of 4.3 U to 1R, only two, involving 18 seedlings, have given simply upright stamens; consequently it may safely be said that no variety has proved pure for upright stamens. In the remaining crosses of this class the ratios have ranged from 1U to 2R up to 10U to 1R, with the greatest frequency at 2U to 1R. The results with the crossed seedlings are practically the same. Over a thousand seedlings from crosses of one type would be expected to give some rather definite results; yet these results are anything but definite, and apparently no conclusions can be drawn from them except that the varieties are not homozygous for uprightness of stamens.

The ratio of practically 1 to 1 in crosses of varieties with reflexed stamens is perhaps best accounted for by the supposition that the gametic composition of pollen and ovules is not alike. The ratio of 1 to 1 in crosses of reflexed by upright stamens may be covered by the same assumption. It should be noted that the pollen of the upright varieties produces the same ratio as that of the reflexed varieties when both are used on ovules of the reflexed kinds.

Upright varieties have been crossed many times with reflexed sorts and several hundred seedlings should have resulted from these, yet only one plant has survived the vicissitudes of the seed bed and nursery to be planted in the test vineyards. In the last five years 50 crosses have yielded 600 seeds; yet from these there are now in the nurseries but 25 living seedlings. Many of the pollen parents used in these crosses were the same as those used in the cross R × R. In the two crosses R × R and R × U, the pollen from upright and reflexed varieties produced the same results; but comparing this last case, U × R, with the first one, U × U, we see that the pollen of the upright and reflexed varieties has produced quite different results when used on upright sorts. Why this should be is not apparent.

At present there does not seem to be any way of eliminating reflexed stamens, but we can at least decrease the proportion by using for breeding only varieties with upright stamens.

INHERITANCE OF SEX.

Among more than 6,000 seedlings which have flowered in the New York Experiment Station vineyards, less than 100 pure male vines

---

*The pollen parent is always placed last.

have been found. Of these there are complete breeding records for 62 vines, 51 of which came from crosses in which the pollen parent was a pure male, leaving 11 males recorded as produced by pollen from hermaphrodite plants. Of these, 5 were pure seedlings from one parent, the other 6 from 5 crosses. These 6 were probably hermaphrodites erroneously recorded as males, an error very easy to make when the pistil has a short style and one which has been made several times and corrected by subsequent observations. The parent yielding the 5 males was discarded shortly after being used in breeding, and our records are meager. It was probably an intermediate recorded as a hermaphrodite. Such an intermediate, having both male and hermaphrodite blossoms, is under observation in one of the Station vineyards, and its pollen seems to behave as the pollen of a pure male; in other words, it is reasonable to assume that, excluding these intermediate forms, pollen from hermaphrodite plants will not produce pure males.

The results obtained from pure males as pollen parents are:

Hermaphrodite female × pure male = 56 hermaphrodites + 51 pure males

Following the assumptions usual to such cases, the hermaphrodite would be considered a sex heterozygote and the male a sex homozygote. Yet selfed hermaphrodites yield only hermaphrodites. These results are similar to those obtained at this Station from selfed hermaphrodite strawberries, but differ from Shull's results (7) with species of Lychnis, where the hermaphrodite gave both females and hermaphrodites. This condition might be covered by the assumption that the hermaphrodite is a female in which the addition of a single dose of maleness has caused the production of male organs, the ovules keeping the composition ♀ ♀ and the pollen becoming ♂ ♀ :

Hermaphrodite × hermaphrodite = ♀ ♀ × ♂ ♀ = 2 ♂ ♀ + 2 ♀ ♀

Since we have no pure females, we must assume that some condition prevents the formation of individuals with the composition ♀ ♀ ; therefore, the above cross gives only hermaphrodites. Of course, if we do not attempt to assume the method of origin of the hermaphrodite, the case may be covered by considering the hermaphrodites pure for this character, while the males would be heterozygous:

$$\text{♀ ♀ × ♀ ♀ = ♀ ♀}$$
$$\text{♀ ♀ × ♀ ♂ = ♀ ♀ + ♀ ♂}$$

COLOR OF SKIN.

Colors of grapes are not sharply differentiated, grading from white through many shades of red and purple to black. Because of this wide range, the problem of finding varieties which are pure for certain colors has been greatly complicated, yet until we are able to find such pure colors and to study their various combinations our knowledge of the color composition of many varieties will probably be only conjectural.

The thousands of seedlings which have been fruited have made possible the formulation of but two general laws: (1) White is a pure color; and (2) white is recessive to both black and red. White, yellow, green, and amber are all considered under the one term and are regarded as being the absence of red and black.

No black variety that has been tested to an extent sufficient to make the results at all conclusive has proved pure for blackness. Some have a factor for red; others seem to contain only black and its absence, white, while still others have both white and red progeny.

In order to simplify the study of red varieties, it has seemed best to divide the seedlings into two shades, the light or medium reds and those ranging from dark red to purple. Those in the second classification are probably red plus either an intensifying factor or various amounts of black. Such a division is somewhat arbitrary and some colors are difficult to place, but, in general, it is a helpful arrangement. Table I gives the result of combining similar colors and shows that results obtained from crosses of red varieties are as diverse as those from crosses of blacks. The table includes mainly pure seedlings.

TABLE I.— RESULTS OF CROSSING GRAPES OF SIMILAR COLORS.

	COLORS OF SEEDLINGS.			
COLORS OF PARENTAL TYPES.	Black.	Purple to dark red.	Medium to light red.	White.
White × white...............	....	....	....	166
Light red × light red...............	8	6	13	8
Dark red × dark red...............	38	43	45	42
Black × black...................	407	49	13	54

Table II illustrates the variation in color composition among most of the varieties given above in the cross black × black and shows the number of varieties which fall in similar groups. From this table it will be seen that 15 black varieties have given only black seedlings, but the number of seedlings is not large enough to be conclusive.

TABLE II.— COLOR GROUPS OF PURE SEEDLINGS OF BLACK VARIETIES OF GRAPES.

NUMBER OF PARENTAL VARIETIES.	COLORS OF SEEDLINGS.		
	Black.	Red.	White.
6	52	16	29
6	128	31	....
10	71	....	25
15	132	....	....

It is interesting to note that the parents with only black and white seedlings produce these colors in the ratio of 3 to 1 and that the seedlings of the varieties yielding only black and red are reasonably close to this same ratio.

The results when different colors are combined are given in Table III. The range of color in the seedlings again emphasizes the necessity of knowing more of the color composition of a variety than can be determined from its appearance.

TABLE III.— RESULTS OF CROSSING GRAPES OF DIFFERENT COLORS.

COLOR COMBINATIONS OF PARENTS.*	COLORS OF SEEDLINGS.			
	Black.	Purple to dark red.	Medium to light red.	White.
White × dark red †	5	44	14	50
White × black	41	3	3	12
Black × dark red	100	52	40	32

* Light-red varieties were not used to an extent sufficient to make the results of value

† The reciprocal cross is included in each case.

It would take up altogether too much space to report upon the color of the progeny of all the varieties studied, but a few of the more common ones are given in Table IV, in order to show the wide variation in different varieties of similar color.

TABLE IV.—VARIATION IN COLOR OF PURE SEEDLINGS OF CERTAIN VARIETIES OF GRAPES.

Name of variety.	Color.	Black.	Purple to dark red.	Medium to light red.	White.
Agawam.....	Purple red..............	1	2	....	2
Brighton.....	Dark red..............	6	5	9	7
Catawba...:.	Purple red..............	2	4	3	4
Champion....	Black..............	13	1	1	2
Clinton......	Black..............	15	....	....	7
Concord.....	Black..............	40	6	....	12
Essex........	Purple black to black....	4	2	3	....
Hartford.....	Black..............	4	....	1	3
Hercules.....	Black..............	3	1	1	10
Isabella......	Black..............	8	....	....	1
Merrimac....	Black..............	9	3	6	....
Nectar.......	Black..............	4	5	2	....
Ozark.......	Black..............	16	....	....	....
Pearl........	White..............	......	....	....	15
Regal........	Dark red.............	......	....	15	5
Worden......	Black..............	4	3	....	1
Wyoming....	Dark red..............	1	4	2	3

In the ultimate solution of the problems of color inheritance we shall probably be aided in no small degree by those who are studying the subject from the standpoint of the chemistry of the various colors; thus, Wheldale (8) has isolated two anthocyanins from species of Antirrhinum which produce different shades of red and three flavones for ivory, yellow, and white. Some work has already been done along this line with the grape. Dezani (3) has found two chromogenic substances in white grapes, and several have reported work on the coloring matter of red grapes, but apparently the results are as yet too indefinite to be of much value to the breeder.

INHERITANCE OF QUALITY.

At first thought it would seem useless to attempt a study of such an elusive and composite character as quality the interpretation of which depends so much upon the tastes of the observers; yet in the final analysis it is this character which very largely determines whether a seedling is worth saving or must go to the brush pile, and any addition to our knowledge of its inheritance is worth the effort.

Table V shows the rating of the progeny of various parental combinations which run the gamut of quality. Most noticeable is the very low percentage of seedlings whose quality is good or above good even when parents of the highest quality were used. When we consider the ancestral history of these seedlings, these results are not surprising or discouraging. Our American grapes, except for the *V. vinifera* hybrids, are but a step removed from the wild, only a few possessing sufficient quality to make them stand out from the many thousands too poor to be eaten with relish. In breeding from these we are breeding from the topmost point of the species and the effect of the several hundred poor kinds in the immediate ancestry is to pull the seedlings down toward the "level of mediocrity."

TABLE V.— INHERITANCE OF QUALITY IN GRAPES.

PARENTAL TYPES.	Best.	Very good to best.	Very good.	Good to very good.	Good.	Total good to best.	Fair to good.	Fair.	Medium.	Poor.	Total, poor to best.	Percentage of good or better.
Very good to best × very good	....	.....	....	2	9	11	4	7	....	1	23	48
Very good × very good	....	1	7	5	22	35	8	24	3	14	84	41
Very good × good to very good	....	......	2	8	30	40	9	30	2	20	101	40
Very good × good	....	......	....	3	6	9	3	43	7	23	85	10
Good to very good × good to very good	....	......	1	6	43	50	20	48	6	36	160	31
Good to very good × good	....	......	1	......	2	3	8	38	2	15	66	4
Good × good	....	1	5	3	31	40	18	59	15	31	163	24
Good × fair	....	......	....	1	3	4	1	6	1	4	16	25
Fair to good × poor	....	......	......	......	2	2	5	9	1	23	40	5
Medium × medium	....	......	1	3	18	22	14	54	20	103	213	10
Poor × poor	....	......	....	1	1	2	....	7	2	40	51	4
Total number of progeny	....	2	17	32	167	218	90	325	59	310	1,002	21

The tendency for the proportion of seedlings of good quality to decrease as we use parents of poorer quality shows clearly the importance of breeding from varieties of only the highest excellence, and even then we must be reconciled to a relatively small percentage of seedlings of good quality.

Practically every grape in the vineyards of the New York Station which ranks high in quality possesses some blood of *V. vinifera*. A moment's consideration of the history of the species shows us the reason for this predominance. European grapes are centuries removed from the wild and have been subjected to a more intense selection than any other fruit; the "level of mediocrity" has been raised to such a point that the species has become a powerful factor in transmitting high quality.

In this connection it is well to speak again of the future that lies ahead of the breeder who will search out and use those varieties of this potent species which blend best with our hardy native kinds. The ages of selection and breeding in Europe have developed varieties of this one species adapted to nearly as wide a range of climate as is covered by all our native species taken together. The proper selection of parents among these should enable us greatly to extend and enrich our viticulture.

A considerable proportion of the seedlings the results of whose crossing are given in Table V are pure seedlings. These have been separated and tabulated in Table VI. Comparing the percentage of those pure seedlings which are good or above good in quality with the percentage of the remaining similar cross-bred seedlings shows an interesting condition. The pure seedlings are uniformly poorer in quality than the crossed seedlings. Is this due to the decrease in vigor which seems to follow selfing, or is there some weightier reason?

TABLE VI.— QUALITY OF PURE SEEDLINGS OF GRAPES.

PARENTAL TYPES.	TYPES OF PROGENY.				
	Good or above good.	Below good.	Total.	Percentage good or above.	Percentage of crossed seedlings good or above.
Very good × very good........	14	25	39	36	47
Good to very good × good to very good..................	6	26	32	19	34
Good × good..................	33	102	135	24	25
Medium × medium...........	17	147	164	10	17
Poor × poor..................	0	31	31	0	10
Total..................	70	331	401	17	31

SIZE OF BERRY.

In order to economize space it was necessary to plant the seedlings so close together in the test vineyards that the clusters frequently did not reach full and characteristic size. For this reason the size of cluster can not be discussed, although it is an important factor. The size of the berry, on the other hand, is one of the size factors least influenced by environment and season. The data from these vines should be of value and are presented in Table VII.

TABLE VII.— INHERITANCE OF THE SIZE OF THE GRAPE BERRY.

PARENTAL TYPES.	CLASSES OF SEEDLINGS.						
	Very small.	Small.	Below medium.	Medium.	Above medium.	Large.	Very large.
Large × large................	1	2	3	*28	*28	19	8
Large† × medium to large‡..	1	6	7	56	34	*67	6
Large × medium.............	......	1	......	*4	2	2	....
Medium to large × medium to large..................	5	34	35	*103	59	20	4
Medium to large × medium..	......	20	35	*57	37	3	2
Medium × medium..........	4	49	39	*83	38	11	....
Medium to large × small....	......	*13	11	12	12	3	....
Medium to small × medium to small.................	26	*35	12	23	7	......	....
Small × small..............	5	*16	4	5	3	......	....

* Numbers in the bold-face type represent the mode.
† The reciprocal is included in each cross.
‡ The use of the two terms shows that the berries varied from medium to large in the same variety.

A study of the various crosses which have entered into Table VII has failed to show any indication of purity for size among the varieties studied. Lacking exact measurements for the various sizes, it is not possible to compute an accurate mean, but the relative position of the mean with respect to the mode can be determined by a short study of the table. The wide variation about the mean, even in crosses where both parents were of the same size, prevents the only cross made between extremes of size, medium to large × small, from showing any clear tendency for the F_1 progeny to be intermediate. The steady decrease in the mean and mode as the parental types grow smaller shows clearly the strong tendency for a variety to produce progeny centering around its own size.

FORM OF BERRY.

The ovalness of many varieties of *V. vinifera* is so pronounced that some have given this as a species characteristic and have assumed that ovalness in our American grapes was an indication of the presence of blood of this species, an assumption hardly warranted by the facts. The large number of markedly oval varieties among table varieties of *V. vinifera*, together with the complete or nearly complete loss of this extreme form in hybrids with our American grapes, would lead us to suppose that this pronounced ovalness is perhaps a nearly pure form and that it is either recessive to roundness or else unites with roundness to produce a less pronounced oval. It is this latter type of oval that is referred to in Table VIII showing the inheritance of berry form. The appearance of so many seedlings with round berries in crosses of such oval varieties would tend to strengthen the idea that this is an intermediate form.

Any study of oblateness is made uncertain by the small number of varieties that possess this form. One of the most pronounced is the Goff, a seedling originated at this Station. The behavior of pure seedlings of the Goff grape would seem to indicate that, in this variety at least, oblateness is a pure form and its disappearance when combined with round, as is shown in Table VIII, would seem to show it as recessive to round.

TABLE VIII.— INHERITANCE OF FORM OF THE GRAPE BERRY.

PARENTAL TYPES.	TYPES OF PROGENY.						
	Oblate.	Slightly oblate.	Oblate to round.	Round.	Round to oval.	Slightly oval.	Oval.
Oval × oval		1	1	*15	7	2	11
Oval × round to oval				*56	10	15	10
Round to oval × round to oval	3	3	10	*129	30	25	56
Round × oval				*15	1	6	3
Round × round to oval		1	2	*100	9	14	1
Round × round	10	17	22	*333	34	24	17
Round to oval × round to oblate	3		1	*17	2	2	1
Round × round to oblate		4	2	*42	5	2	1
Round to oblate × round to oblate	3		1	*24	2	1	8
Round × oblate				7			
Oblate × oblate	*15			1			

* Numbers in bold-face type represent the mode.

From a study of Table VIII it is seen that the mean would be more nearly coincident with the mode in each cross than was the case in Table VII. This shows clearly the strong tendency for roundness to obscure both oval and oblate.

SEASON OF RIPENING.

The period of ripening of a variety depends so much upon the vigor of the vine, the season, cultural methods, and environmental conditions that no very accurate data can be presented. In one year all varieties may be 10 days earlier than normally, while in another year early varieties may be unusually early; but a cold, wet period late in September and early in October may cause the late varieties to be unusually late. These variations are minimized when the records extend over a number of years. The ripening dates of the seedlings are usually taken for at least three years — not long enough, but much better than if taken for a single year.

In Table IX the ripening season extends approximately through the months of September and October. The first two periods cover about 15 days each, the next two about 10 days each, while the length of the last period is usually fixed by the first killing frosts.

TABLE IX.— EFFECT OF HEREDITY ON SEASON OF RIPENING OF GRAPES.

PARENTAL TYPES WITH REFERENCE TO RIPENING SEASON.	RIPENING PERIODS OF PROGENY.					
	Approximate mean.	Very early.	Early.	Early midseason.	Midseason.	Late.
Early × early	Sept. 23	8	*30	20	......	.....
Early × early to midseason †	Sept. 22	13	*46	18	......	.....
Early to midseason × early to midseason	Sept. 27	7	*46	42	7	1
Early × midseason	Sept. 28	......	20	*22	......	.....
Early to midseason × midseason	Sept. 26	21	*126	100	8	.....
Midseason × midseason	Oct. 1	11	165	*244	49	.....
Early × late	Oct. 1	......	8	8	......	.....
Early to midseason × late	Sept. 27	2	*9	2	......	3
Midseason × late	Oct. 4	3	20	*27	18	6
Late × late	Oct. 7	......	10	*104	19	14

* Numbers in boldface type represent the mode.
† Each cross includes also the reciprocal.

As would be expected, Table IX fails to show purity or dominance for any one season, but it does show, both in the mode and in the approximate mean, the extent to which the season of the parent influences the offspring. A study of the varieties which enter into the table has failed to show results at all different from those of the group in which they fall.

NEW VARIETIES FROM EARLIER CROSSES.

The results of the first 20 years of work were anything but encouraging. Now, however, there is tangible evidence that progress is being made. A vineyard of 1,500 seedlings bred from 1898 to 1903 has by a process of vigorous selection decreased to less than 75 vines, but among this number are several that seem very promising. Five of these have already proved so desirable both at Geneva and in a test vineyard at the Station's Vineyard Laboratory at Fredonia, N. Y., that in the fall of 1914 it was decided to give them names and place them in the hands of the nurserymen.

LITERATURE CITED.

(1) BEACH, S. A.
   1898. Self-fertility of the grape. N. Y. State Agr. Exp. Sta. Bul. 157, p. 397–441, 3 fig., 5. pl.
(2) BOOTH, N. O.
   1902. Investigations concerning the self-fertility of the grape, 1900–1902. III. A study of grape pollen. N. Y. State Agr. Exp. Sta. Bul. 224, p. 291–302, 1 fig., 6 pl.
(3) DEZANI, Serafino.
   1910. Le sostanze cromogene dell' uva bianca. *Staz. Sper. Agr. Ital.*, 43:428–438. Abstract in *Jour. Chem. Soc.* [London], 100:223. 1911.
(4) DORSEY, M. J.
   1914. Pollen development in the grape with special reference to sterility. Minn. Agr. Exp. Sta. Bul. 144, 60 p., 4 pl. Bibliography, p. 50–60.
(5) ———
   1914. Sterility in the grape. Proc. Soc. Hort. Sci., 1913, 10th Ann. Meeting, p. 149–153.
(6) REIMER, F. C., and DETJEN, L. R.
   1914. Breeding rotundifolia grapes. N. C. Agr. Exp. Sta. Tech. Bul. 10, 47 p., 19 fig.
(7) SHULL, G. H.
   1911. Reversible sex-mutants in Lychnis dioica. *Bot. Gaz.*, 52:329–368, 15 fig. Literature, p. 366–368.
(8) WHELDALE, M.
   1914. Our present knowledge of the chemistry of the Mendelian factors for flower-colour. *Jour. Genetics*, 4:109–129, 5 fig., pl. 7 (col.). Bibliography, p. 127–129.

Lightning Source UK Ltd.
Milton Keynes UK
UKHW020401081118
331957UK00009B/816/P